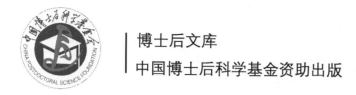

博士后文库
中国博士后科学基金资助出版

超冷原子系统中隧穿动力学的量子操控

豆福全 著

科学出版社
北京

内 容 简 介

本书主要介绍超冷原子系统中隧穿动力学量子操控领域的几个重要内容,包括超冷原子系统的非线性隧穿动力学、高保真度量子操控和超冷原子-分子转化动力学,研究粒子间相互作用等非线性因素导致的新奇量子现象,探索一些新颖的技术手段. 全书共 5 章,第 1、2 章主要介绍研究背景、研究意义以及相关概念,第 3~5 章主要介绍超冷原子系统隧穿动力学量子操控研究中的三个重要内容.

本书适合物理学、力学、数学以及化学等专业的有关教师和研究生阅读,也可供相关专业的科研人员参考.

图书在版编目(CIP)数据

超冷原子系统中隧穿动力学的量子操控/豆福全著. —北京:科学出版社,2019.9
(博士后文库)
ISBN 978-7-03-062298-3

Ⅰ. ①超⋯ Ⅱ. ①豆⋯ Ⅲ. ①原子物理学–研究 Ⅳ. ①O562

中国版本图书馆 CIP 数据核字 (2019) 第 204585 号

责任编辑:宋无汗 崔慧娴/责任校对:郭瑞芝
责任印制:师艳茹/封面设计:陈 敬

科学出版社 出版
北京东黄城根北街 16 号
邮政编码:100717
http://www.sciencep.com
北京凌奇印刷有限责任公司 印刷
科学出版社发行 各地新华书店经销
*
2019 年 9 月第 一 版 开本:720×1000 B5
2019 年 9 月第一次印刷 印张:12
字数:242 000
POD定价:98.00元
(如有印装质量问题,我社负责调换)

《博士后文库》编委会名单

主　任　陈宜瑜

副主任　詹文龙　李　扬

秘书长　邱春雷

编　委　（按姓氏汉语拼音排序）

付小兵　傅伯杰　郭坤宇　胡　滨　贾国柱　刘　伟

卢秉恒　毛大立　权良柱　任南琪　万国华　王光谦

吴硕贤　杨宝峰　印遇龙　喻树迅　张文栋　赵　路

赵晓哲　钟登华　周宪梁

《博士后文库》序言

1985 年，在李政道先生的倡议和邓小平同志的亲自关怀下，我国建立了博士后制度，同时设立了博士后科学基金. 30 多年来，在党和国家的高度重视下，在社会各方面的关心和支持下，博士后制度为我国培养了一大批青年高层次创新人才. 在这一过程中，博士后科学基金发挥了不可替代的独特作用.

博士后科学基金是中国特色博士后制度的重要组成部分，专门用于资助博士后研究人员开展创新探索. 博士后科学基金的资助，对正处于独立科研生涯起步阶段的博士后研究人员来说，适逢其时，有利于培养他们独立的科研人格、在选题方面的竞争意识以及负责的精神，是他们独立从事科研工作的"第一桶金". 尽管博士后科学基金资助金额不大，但对博士后青年创新人才的培养和激励作用不可估量. 四两拨千斤，博士后科学基金有效地推动了博士后研究人员迅速成长为高水平的研究人才，"小基金发挥了大作用".

在博士后科学基金的资助下，博士后研究人员的优秀学术成果不断涌现. 2013 年，为提高博士后科学基金的资助效益，中国博士后科学基金会联合科学出版社开展了博士后优秀学术专著出版资助工作，通过专家评审遴选出优秀的博士后学术著作，收入《博士后文库》，由博士后科学基金资助、科学出版社出版. 我们希望，借此打造专属于博士后学术创新的旗舰图书品牌，激励博士后研究人员潜心科研，扎实治学，提升博士后优秀学术成果的社会影响力.

2015 年，国务院办公厅印发了《关于改革完善博士后制度的意见》（国办发〔2015〕87 号），将"实施自然科学、人文社会科学优秀博士后论著出版支持计划"作为"十三五"期间博士后工作的重要内容和提升博士后研究人员培养质量的重要手段，这更加凸显了出版资助工作的意义. 我相信，我们提供的这个出版资助平台将对博士后研究人员激发创新智慧、凝聚创新力量发挥独特的作用，促使博士后研究人员的创新成果更好地服务于创新驱动发展战略和创新型国家的建设.

祝愿广大博士后研究人员在博士后科学基金的资助下早日成长为栋梁之才，为实现中华民族伟大复兴的中国梦做出更大的贡献.

中国博士后科学基金会理事长

作者简介

豆福全：北京理工大学博士，北京应用物理与计算数学研究所博士后．主要研究方向为冷原子物理、非线性物理和相关量子调控问题及应用．目前主持国家自然科学基金 2 项，甘肃省自然科学基金 1 项，完成中国博士后科学基金面上项目(一等资助) 1 项，参与国家自然科学基金重大研究项目等 6 项．发表 SCI 等论文 50 余篇．获甘肃省高校科技进步奖一等奖 2 项、二等奖 4 项，甘肃省高等学校科研优秀成果奖科学技术类三等奖 2 项．

前 言

超冷原子系统的量子操控是原子分子和光物理等领域近年来最重要、最活跃、发展最为迅速、成果最为辉煌的前沿课题之一. 这一前沿研究和化学、量子信息、凝聚态物理甚至天体物理紧密相关, 会出现很多超乎寻常的量子现象, 也可能带来一系列新的技术突破. 这些研究既有很强的基础理论价值, 也有广泛的应用前景. 本书围绕超冷原子系统隧穿动力学量子操控这一前沿领域, 主要介绍超冷原子系统的非线性隧穿动力学、高保真度量子操控和超冷原子–分子转化动力学, 研究粒子间相互作用等非线性因素导致的新奇量子现象, 探索量子操控的一些新颖的技术手段.

全书共 5 章, 第 1、2 章主要介绍研究背景、研究意义以及相关概念, 第 3~5 章介绍超冷原子系统隧穿动力学量子操控研究中的三个重要内容.

第 1 章对本书的研究背景做了简要介绍. 主要包括超冷原子物理、量子相干操控、超冷原子系统量子操控的研究意义.

第 2 章介绍本书所述的超冷原子系统隧穿动力学量子操控研究中三个方面的研究现状、基本方法以及研究意义等, 包括非线性隧穿动力学、高保真度量子操控、超冷原子–分子转化.

第 3 章从数值和解析两方面研究具有粒子间相互作用和非线性扫描两种非线性的两能级系统的 Landau-Zener 隧穿动力学, 以及具有粒子间相互作用的非线性 Demkov-Kunike 跃迁动力学. 在 Landau-Zener 隧穿中, 由于两种非线性的相互竞争, 在不同参数范围内, 发生一系列有趣的现象, 如绝热性的破坏、干涉现象的消失以及隧穿概率的不对称性. 在 Demkov-Kunike 跃迁动力学中, 由于非线性粒子间相互作用的存在, 跃迁动力学中也出现了不对称性等. 同时, 还介绍一些实现这些隧穿动力学的可能实验方案.

第 4 章介绍两能级系统高保真度量子驱动问题, 包括超绝热量子驱动、超快量子驱动以及复合绝热通道技术. 在超绝热量子驱动问题中, 以啁啾高斯两能级模型和 Demkov-Kunike 模型为例, 发现通过构造和调节附加哈密顿量反非绝热场 (counter-diabatic field, CD 场), 非绝热损失能被抵消, 且精度达到任意阶, 即使在原来的非绝热参数范围也能保证绝热跟随瞬时绝热基态, 实现一个高保真度、鲁棒且加速的无跃迁超绝热布居数转移. 在超快量子驱动问题中, 主要考虑非线性两能级系统, 一是具有粒子间相互作用的非线性两能级系统, 通过与非线性 Landau-Zener 和 Roland-Cerf 技术比较, 发现复合脉冲 (composite pulse) 技术能实现从初态到目标态的最大量子转化速度, 探讨原子间相互作用对高保真度量子驱动的影响, 发现系统中存在一种粒子间相互作用的临界行为; 二是进一步考虑具有粒子间相互作用和非线性扫描两种非线性的广义两能级模型, 系统研究超快量子控制问题, 也讨论模型的实验方案以及在加速光晶格玻色–爱因斯坦凝聚中观察到该非线性效应的可能性. 在复合绝热通道的问题中, 将复合绝热通道技术推广到有限时间的两能级系统和非线性两能级系统中, 实现快速、高保真度量子控制.

第 5 章介绍超冷原子–分子的转化问题. 一是考虑包含粒子间相互作用的超冷玻色系统中原子–双原子分子的转化, 通过在 Feshbach 共振附近设计一个磁场脉冲链, 实现一种稳定、高效产生超冷双原子分子的方法. 得到这些磁场脉冲持续时间的解析表达式, 研究分子转化效率和每个磁场脉冲的持续时间的依赖关系, 发现只要适当地调节各个磁场脉冲的持续时间, 分子转化效率可高达 100%, 也讨论了粒子间相互作用对原子–分子转化过程的影响. 二是介绍超冷玻色系统中同核和异核原子–多聚物分子的转化问题. 首先考虑超冷五聚物分子的形成, 接着对该方法进行拓展, 研究超冷 N 体 Efimov 多聚物分子的产生. 利用一种广义的受激拉曼绝热通道技术, 建立平均场模型, 获得系统的 CPT 解, 对多聚物分子, 得到其 CPT 解满足一个普适的代数方程. 利用线性不稳定分析方法, 分析暗态的稳定性, 并通过定义合适的保真度研究系统的绝热性. 最后讨论粒子间相互作用、多聚物分子中原子数目以及外场参数、外场形式对原子–多聚物分子转化过程的影响.

特别感谢刘杰研究员、傅立斌研究员多年来的指导、鼓励和帮助, 感谢团队的各位成员, 许多问题是在和大家的讨论中得以解决, 感谢各位老师、同学、同事、朋友和家人长期以来的指导、鼓励和支持. 感谢国家自然科学基金 (11665020, 11547046) 和中国博士后科学基金对作者课题的资助, 感谢中国博士后科学基金优秀学术专著出版资助基金对本书的资助. 感谢科学出版社各位编辑对本书出版所付出的辛勤努力.

目前, 关于超冷原子系统量子操控方面的研究正处于蓬勃发展时期, 该研究涉及学科方向众多, 和许多学科高度交叉, 新的技术方法层出不穷, 新的研究成果不断涌现, 具有极为广阔的应用前景. 国内的许多科研院所及高校也取得了一系列杰出的成果. 本书主要介绍超冷原子系统非线性隧穿动力学、高保真度量子操控以及超冷原子–分子转化, 所涉范围相对较窄, 加之作者水平和能力有限, 书中疏漏和不当之处在所难免, 敬请批评指正.

目　　录

《博士后文库》序言
前言
第 1 章　绪论：超冷原子物理——极热研究领域 ·················· 1
　1.1　超冷原子物理 ·· 1
　　　1.1.1　低温物理 ·· 1
　　　1.1.2　玻色–爱因斯坦凝聚 ·· 4
　　　1.1.3　超冷原子性质 ··· 6
　1.2　量子相干操控 ·· 7
　　　1.2.1　研究背景与研究意义 ·· 7
　　　1.2.2　量子相干操控概念 ··· 8
　　　1.2.3　超冷原子系统中量子操控的方法 ······························· 8
　1.3　超冷原子系统量子操控的研究意义 ····································· 10
　参考文献 ·· 14
第 2 章　隧穿动力学：量子操控重要方向 ······························ 18
　2.1　非线性隧穿动力学 ··· 18
　　　2.1.1　基本模型 ·· 18
　　　2.1.2　几种常见的隧穿模型 ··· 19
　　　2.1.3　非线性方面的扩展 ·· 21
　2.2　高保真度量子操控 ··· 24
　　　2.2.1　超绝热量子操控 ··· 24
　　　2.2.2　超快量子操控 ·· 25
　　　2.2.3　复合绝热通道技术 ·· 26
　2.3　超冷原子–分子转化 ··· 27
　　　2.3.1　超冷分子性质 ·· 27
　　　2.3.2　超冷分子的几种产生方法 ······································· 28
　　　2.3.3　Efimov 共振及其超冷多聚物分子的形成 ···················· 39
　参考文献 ·· 45
第 3 章　超冷原子系统中的非线性隧穿动力学 ······················· 58
　3.1　广义 Landau-Zener 隧穿 ··· 58

3.1.1 粒子间相互作用和非线性扫描对 LZT 的影响 ········· 59
3.1.2 应用举例 ······························· 65
3.2 非线性 Demkov-Kunike 跃迁 ························ 66
3.2.1 模型 ································· 66
3.2.2 无静态失谐情况 ·························· 67
3.2.3 存在静态失谐情况 ························· 73
3.2.4 应用和讨论 ····························· 75
参考文献 ······································ 76

第 4 章 高保真度量子操控 ···························· 81
4.1 高保真度超绝热量子驱动 ·························· 81
4.1.1 模型和超绝热技术 ························· 82
4.1.2 啁啾 Gaussian 模型中的高保真度超绝热量子驱动 ······· 84
4.1.3 Demkov-Kunike 模型的高保真度超绝热量子驱动 ······· 89
4.1.4 结论与讨论 ····························· 97
4.2 高保真度超快量子驱动 ··························· 97
4.2.1 具有粒子间相互作用两能级系统中的超快量子驱动 ······ 97
4.2.2 广义非线性两能级系统中高保真度超快量子驱动 ······· 104
4.3 高保真度复合绝热通道技术 ······················· 112
4.3.1 复合绝热通道技术 ························ 114
4.3.2 有限时间两能级系统中高保真度布居数转移 ········· 115
4.3.3 非线性两能级系统中的应用 ··················· 120
4.3.4 结论与讨论 ····························· 125
参考文献 ······································ 126

第 5 章 超冷原子-分子转化动力学 ······················ 134
5.1 超冷双原子分子的产生: 磁场脉冲链技术的量子操控 ········· 134
5.1.1 平均场模型和拉比振荡 ······················ 135
5.1.2 原子-分子转化的磁场脉冲链技术 ················ 138
5.1.3 结论和讨论 ····························· 142
5.2 超冷 N 体 Efimov 多聚物分子的形成: 广义受激拉曼绝热通道
   技术的量子操控 ····························· 143
5.2.1 超冷五聚物分子的形成 ······················ 145
5.2.2 超冷 N 体 Efimov 多聚物分子的形成 ············· 153
5.2.3 小结 ································· 162
5.3 外场形式对超冷原子-多聚物分子转化效率的影响 ·········· 162

5.3.1 模型与绝热保真度 ································· 163
　　5.3.2 外场扫描形式对转化效率的影响 ···················· 164
　　5.3.3 小结 ··· 169
参考文献 ··· 169
编后记 ··· 176

第1章 绪论:超冷原子物理——极热研究领域

本书涉及"超冷原子量子操控"的几个方面:超冷原子系统中的非线性隧穿动力学、高保真度量子操控以及超冷原子–分子转化. 本章主要介绍与该领域相关的基本概念、研究背景以及研究意义等,包括超冷原子物理、量子相干操控以及超冷原子系统量子操控的研究意义.

1.1 超冷原子物理

当您阅读本书时,空气分子正以大约 500 m/s 的平均速度从您身旁穿梭而过(这个速度比一般手枪子弹的速度要快),而且从四面八方袭来. 与此同时,构成人体的原子和分子也正不停翻滚、振动或互相碰撞着. 物体的行进速度越快,意味着蕴含的能量就越大. 原子和分子的集体能量,通常就是人们所感受到的热,而温度正是描述物质内部热运动量度的物理量. 自然界中没有任何东西是完全静止的,如果所有的原子都停下躁动的脚步,静止不动,就意味着温度为绝对零度 (−273.15℃, 19 世纪中叶由开尔文定义). 热力学第三定律也告诉人们,绝对零度不可能达到. 然而,低温世界就像魔术师,会使物质世界出现各种奇异现象,因此人们对低温世界的探索永不止步. 目前,随着人类对世界认识的不断深入以及对微观操控和冷却技术的不断提高,物质世界的温度也在不断降低,已经越来越无限地接近绝对零度,低温物理便很自然地成为物理学大厦中很重要的一个分支.

1.1.1 低温物理

低温物理是物理学工作者极其关注的重要物理领域之一. 人类对世界的认识也随着温度的降低而不断深入,而且每次低温的突破几乎都会和诺贝尔奖结缘.

1911 年 Onnes[1] 关于超导和 1938 年 Kapitza[2] 关于 ^{4}He 超流的发现,使温度环境进入到大约 1K 的范围,Onnes 和 Kapitza 分别获得 1913 年和 1978 年的诺贝尔物理学奖. 1972 年 Osheroff 等 [3] ^{3}He 超流的发现又将温度降低到毫开 (mK) 范围 (多普勒冷却),即 10^{-3}K,因此获得 1996 年的诺贝尔物理学奖. 1985 年 Chu[4], Cohen-Tannoudji[5] 以及 Phillips[6] 等首次利用激光冷却技术将原子温度降低到微开 (μK) 范围 (亚多普勒冷却),即 10^{-6}K,从而获得了 1997 年的诺贝尔物理学奖. 1995 年 JILA、Rice 和 MIT 的实验小组利用激光冷却、原子囚禁和蒸发冷却技术将原子进一步冷却到纳开 (nK) 范围,即 10^{-9}K,相继实现了碱金属原子 ^{87}Rb、^{7}Li 和

^{23}Na 的玻色–爱因斯坦凝聚 (Bose-Einstein condensation, BEC)[7-10], 美国天体物理学联合实验研究所 (Joint Institute for Laboratory Astrophysics, JILA) 的 Cornell、Wieman 与麻省理工学院 (Massachusetts Institute of Technology, MIT) 的 Ketterle 因此共同获得了 2001 年的诺贝尔物理学奖. 此后, 冷原子物理领域又掀起了一次研究热潮, 大量的工作涌现出来. 到目前, 最低温度已经达到皮开 (pK) 量级, 即 10^{-12}K 范围 [11]①. 图 1.1 (图中正三角形表示 ^{3}He 超流, 倒三角形表示 ^{3}He 固体, 菱形表示 ^{3}He 和 ^{4}He 混合, 球形表示铜中的传导电子, 右三角表示 Rb 玻色–爱因斯坦凝聚, 方块表示 Cu、Ag 或 Rh 的核自旋, 左三角表示 Ag 或 Rh 负核自旋温度) 和图 1.2 显示了最低温度的研究发展历史. 这些开创性的工作揭开了超冷原子物理研究的崭新篇章 [12,13].

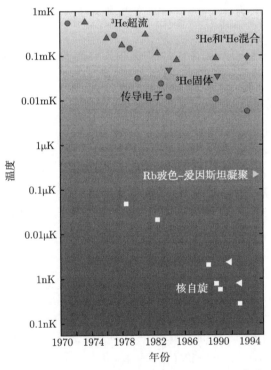

图 1.1 最低温度纪录示意图 (1970~1996) [14]

---

① 据 *Nature* 杂志网站 2013 年 1 月 3 日 (Quantum gas goes below absolute zero, http://www.nature.com/news/ quantum-gas-goes-below-absolute-zero-1.12146), ScienceDaily 网站 2013 年 1 月 4 日 (A temperature below absolute zero: atoms at negative absolute temperature are the hottest systems in the world, http://www.sciencedaily.com/releases/2013/01/130104143516.htm), 以及 *Science* 杂志 2013 年 1 月 4 日报道, 德国物理学家在光晶格中利用钾原子首次造出一种"低于"绝对零度 (负温度) 的量子气体. 实际上这里所说的"负温度"并不表示比绝对零度还低的温度, 可见本书 1.3 节的有关介绍.

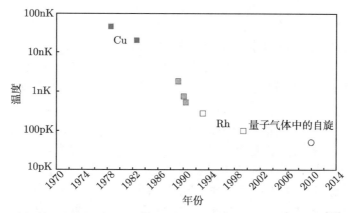

图 1.2　最低温度的研究发展历史示意图 (1970~2014 年) [15]

这里所谓的超冷,就是指原子作为整体的平动速度极低,温度一般低于毫开 (mK) 范围,现在也常常指温度低于微开 (μK) 范围. 如此低的温度,为物理学的环境作了一项革命性的变革. 不同温度的物质,所满足的物理规律也会发生很大的变化. 图 1.3 显示 JILA 实验组对物质温度划分的示意图. 可以看到,温度从开 (K) 到毫开 (mK),再到超冷低于微开 (μK) 范围,所满足的物理规律由经典物理描述到半经典物理,再到完全的量子力学描述. 而玻色–爱因斯坦凝聚的实现就是低温物理研究中的一个杰出代表,它为研究这些超冷环境下的量子现象提供了一个绝佳的工具.

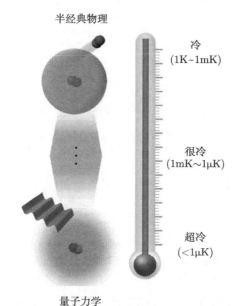

图 1.3　JILA 实验组对物质温度划分示意图 [16]

1.1.2 玻色-爱因斯坦凝聚

冷原子气体中,玻色-爱因斯坦凝聚的实现标志着低温物理中超冷原子气体研究快速发展的开端,已经为物理学提供了一个新的物质态,被称为物质的第五态,可用于进行各种前所未有的研究,涵盖了原子物理、量子光学以及凝聚态物理等众多领域.

气态玻色-爱因斯坦凝聚体是总自旋为整数的气态原子(自然界中的粒子按照自旋分为两种,一种是总自旋为 $\hbar/2$ 偶数倍或 0 的粒子,称为玻色子;另一种总自旋是 $\hbar/2$ 奇数倍的粒子,称为费米子),在一定条件下,尤其是超低温下所呈现的特殊量子状态,会宏观分布在同一状态,即能量最低态. 这是一种宏观量子现象,也就是说,宏观尺度的系统具有了微观客体才有的量子特性. 该现象早在 1924~1925 年由玻色和爱因斯坦预言,并于 1995 年首次在实验中观测到.

1924 年印度物理学家 Bose[17] 在研究黑体辐射谱时得到了光子的统计分布规律(假定光子是不可分辨的). 1925 年,爱因斯坦把这种假定推广到玻色子组成的粒子体系,后来人们就称这种统计规律为玻色-爱因斯坦统计. 该统计表明,在第 i 个量子态上分布的粒子个数为 n_i:

$$n_i = \frac{1}{e^{(\epsilon_i - \mu)/k_B T} - 1}. \tag{1.1}$$

其中,ϵ_i 是第 i 个量子态的能量;μ 是系统的化学势;$k_B = 1.38 \times 10^{23} \mathrm{J \cdot K^{-1}}$ 为玻尔兹曼常量;T 是热力学温度. μ 和 T 受系统的总粒子数 N 和总能量 E 限制:

$$N = \sum_i \frac{1}{e^{(\epsilon_i - \mu)/k_B T} - 1}, \tag{1.2}$$

$$E = \sum_i \frac{\epsilon_i}{e^{(\epsilon_i - \mu)/k_B T} - 1}. \tag{1.3}$$

如果考虑的是一个大系统,上述限制条件可写作积分形式:

$$N = \int d\epsilon \frac{g(\epsilon)}{e^{(\epsilon - \mu)/k_B T} - 1}, \tag{1.4}$$

$$E = \int d\epsilon \frac{\epsilon g(\epsilon)}{e^{(\epsilon - \mu)/k_B T} - 1}, \tag{1.5}$$

其中,$g(\epsilon) = 2\pi V (2m)^{3/2} \epsilon^{1/2}/h^3$ 表示态密度,V 是原子气体体积,m 是原子质量,$h = 6.63 \times 10^{34} \mathrm{J \cdot s}$ 是普朗克常量. 随着温度的降低,系统的化学势($\mu < 0$)逐渐增大. 当温度达到某一临界值 T_c 时,系统的化学势 $\mu \to 0^-$,由上述方程可知,$\epsilon = 0$ 态上的粒子数目可能增大至 N 的量级,甚至整个系统的粒子都有可能集体占据到此量子态上. 这种在绝对零度附近发生的玻色子宏观占据某一量子态的现象即为玻色-爱因斯坦凝聚.

用 $N_0(T)$ 表示温度为 T 时动能为零的原子数, 式 (1.4) 可写作:

$$N = N_0(T) + \int_0^\infty \frac{g(\epsilon)}{e^{(\epsilon-\mu)/k_B T} - 1} d\epsilon = N_0(T) + \frac{V}{\lambda_{\rm dB}^3} g_{\frac{3}{2}}(z), \qquad (1.6)$$

其中, $g_n(z) = \frac{1}{\Gamma(n)} \int_0^\infty \frac{x^{n-1}}{e^x/z - 1} dx$ 为玻色函数, $z = e^{\mu/k_B T}$ 为气体的逸度, $\Gamma(n) = \int_0^\infty e^{-t} t^{n-1} dt$ 为 Gamma 函数; $\lambda_{\rm dB}$ 为热 de Broglie 波长:

$$\lambda_{\rm dB} = \frac{h}{\sqrt{2\pi m k_B T}}, \qquad (1.7)$$

当 $\mu \to 0$ 时, $z \to 1$, $g_{3/2}(1) = 2.612$. 如果进一步降低温度, 增加气体的密度, 由式 (1.6) 的第二个等式可以看出, 此时气体中原子数的增加只能来源于 N_0 这一项, 也就是系统中会有越来越多的原子落到动能为零的态上, 导致玻色-爱因斯坦凝聚的形成. 由此可见, 形成玻色-爱因斯坦凝聚的条件是相空间密度 $\rho_{\rm ps}$ 要满足:

$$\rho_{\rm ps} = n \lambda_{\rm dB}^3 = \frac{N}{V} \lambda_{\rm dB}^3 \geqslant 2.612. \qquad (1.8)$$

这里的相空间密度定义是在自由空间中 $\rho_{\rm ps} = n \lambda_{\rm dB}^3$ 的形式, 它用来衡量样品是否"冷"(在低温且大的粒子数密度时其值大). 由式 (1.8) 可得, 玻色-爱因斯坦凝聚转变温度和临界密度分别为

$$\begin{cases} T_c = \dfrac{h^2}{2\pi k_B} \left(\dfrac{n}{2.612} \right)^{2/3}, \\ n_c = 2.612 \left(\dfrac{2\pi m k_B T}{h^2} \right)^{3/2}. \end{cases} \qquad (1.9)$$

相应地, 可得凝聚体中的原子数随温度的变化:

$$\frac{N_0}{N} = 1 - \left(\frac{T}{T_c} \right)^{3/2}. \qquad (1.10)$$

实验时, 首先利用磁光阱, 然后采用蒸发冷却来实现玻色-爱因斯坦凝聚[7]. 目前已被实验证实能够通过激光冷却和陷俘原子的相关技术产生玻色-爱因斯坦凝聚的原子种类有 ^{87}Rb[7,8]、^{23}Na[10,18]、^{7}Li[9]、^{1}H[19]、^{85}Rb[20]、亚稳的 ^{4}He[21,22]、^{41}K[23]、^{133}Cs[24]、^{174}Yb[25,26]、^{52}Cr[27]、^{39}K[28]、^{170}Yb[29]、^{40}Ca[30]、^{84}Sr[31,32]、^{88}Sr[33]、^{86}Sr[34]、^{168}Er[35] 等. 同时由费米原子对, 如 ^{6}Li 和 ^{40}K 形成的分子玻色-爱因斯坦凝聚也已经实现.

图 1.4 给出了实验中测得的铷原子速度分布随着温度的变化, 其中左边代表温度高于凝聚温度时的气体, 中间表示凝聚体出现, 右边表示经过进一步蒸发冷却后只剩下几乎纯的凝聚体. 颜色对应每个速度下的原子数[7,36]. 当原子气体的温度高于凝聚体转变温度时, 凝聚体没有形成, 原子的分布与热原子平衡态下的分布类似.

温度一旦降低到转变温度以下, 在速度空间对应 $v=0$ 的位置出现一个尖峰, 形成了凝聚体. 随着蒸发冷却的进行, 原子的温度继续降低, 越来越多的原子从非凝聚的状态转移到凝聚体中, 最终形成几乎纯的凝聚体.

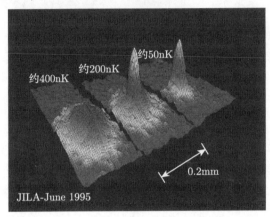

图 1.4　Anderson 等实验中的铷原子速度分布图像 [7, 36]

稀薄量子气体之所以形成玻色-爱因斯坦凝聚, 是因为它们不同于通常的气体、液体和固体. 典型玻色-爱因斯坦凝聚中心的原子团的粒子数密度为 $10^{19} \sim 10^{21}$ 个 $\cdot$ m^{-3}, 而气体分子在室温和标准大气压下的密度是 10^{25} 个 $\cdot$ m^{-3}, 液体和固体原子密度数量级大约为 10^{28} m^{-3}, 而原子核的核子密度大约是 10^{44} 个 $\cdot$ m^{-3}, 可见稀薄气体的密度最低. 要在如此低密度的量子系统中观察到量子现象, 要求系统的温度必须达到 10^{-5} K, 甚至更低. 这和其他系统中观察到量子现象的温度形成了鲜明的对比. 固体中, 要出现明显的量子效应, 在金属中的电子要低于费米温度, 即 $10^4 \sim 10^5$ K. 对声子而言, 温度要低于 Debye 温度, 即 10^2 K 数量级. 对于氦液体, 观察到量子现象的温度数量级是 1K. 而对于原子核而言, 密度更高, 对应的简并温度大约是 10^{11} K [37].

1.1.3　超冷原子性质

原子在常温下会发生无规则热运动, 有些原子在常温下的速度高达每秒数百米, 这对精密测量和操控来说无疑是一种巨大的障碍. 通过 $p=mv=\sqrt{2\pi m k_\mathrm{B} T}$, 即 $v=\sqrt{2\pi k_\mathrm{B} T/m}$ 可以估算原子的速度. 例如, 对于 ^{133}Cs 原子, 常温 ($T=300$K) 下原子的速度约为 $v=343$m/s, 当 $T=1$mK 时, $v=0.627$m/s, 而当温度为 $T=1\mu$K 时, 对应的速度为 $v=0.0198$m/s, 温度低到 $T=1$nK 时, $v=0.000627$m/s. 同样可以利用式 (1.7) 计算 de Broglie 波长, 发现温度越低, 其波长越长. 可以看到, 相对于室温或更高温的热原子而言, "超冷原子" 气体的速度及速度分布宽度要低几个量级. 如此低温下的原子系统, 会出现许多新的奇特现象, 遵从许多新的物理规律,如以下几点[38].

(1) 动能小, 可用弱场, 如磁场梯度、光场操控, 其空间位置可高精度控制, 由此可以精确地控制原子的动力学行为和内部状态.

(2) de Broglie 波长大, 原子的物质波波动性明显.

(3) 热运动小, 光谱一级和二级 Doppler 加宽很小, 也有利于精密测量.

(4) 速度低, 原子与光相互作用时间长.

因此, 超冷原子系统具有精确可控、受外界环境影响小等优点, 这些新的性质为冷原子的应用奠定了坚实的基础.

1.2 量子相干操控

1.2.1 研究背景与研究意义

2012 年 10 月 9 日, 瑞典皇家科学院诺贝尔奖评审委员会宣布将 2012 年度的诺贝尔物理学奖授予法国巴黎高等师范学院的 Haroche 教授和美国国家标准局的 Wineland 教授, 以表彰他们"提出了突破性的实验方法, 使测量和操控单个量子系统成为可能". 过去数十年里, 这两位物理学家主要围绕原子、离子和光子的操控和测量, 在光与物质相互作用最基本的层面上, 即在确定性的单个粒子的水平上, 为人们展示了微观世界中一系列丰富多彩的量子行为. 其最大的贡献在于首次让这一领域的研究向应用层面发展, 使新一代的超级量子计算机的诞生有了初步可能.

事实上, 微观粒子 (原子、分子、离子等) 的量子操控一直是现代物理学研究最重要、最活跃的前沿领域之一, 三十多年来有多次诺贝尔物理学奖授予该领域科学家, 包括离子俘获 (1989)、激光冷却和陷俘原子 (1997)、玻色–爱因斯坦凝聚 (2001)、量子光学和激光光谱学 (2005) 以及测量和操控单个量子系统 (2012), 还有许多诺贝尔物理学奖和化学奖与该领域密切相关, 如整数 (1985) 和分数 (1998) 量子霍尔效应、氦-3 超流的发现 (1996)、高温超导 (1987)、超导和超流理论 (2003)、电子和扫描隧道显微镜 (1986)、量子光学和激光光谱学 (1981、1999 化学)、电子态计算 (1998 化学) 等[39]. 美国科学院联合会发布的原子、分子物理和光学委员会 2010 年的前瞻报告, 标题就是"调控量子世界". 2016 年 7 月, 美国国家科学技术委员会 (National Science and Technology Council, NSTC) 发布《推动量子信息科学: 国家的挑战与机遇》的报告, 再次强调发展量子信息科学的重要性, 总结了量子信息科学在多个领域的进展和未来发展潜力, 包括量子传感与计量、量子通信、量子模拟及量子计算等方面. 英国于 2014 年制定了量子科学五年计划, 我国的《国家中长期科学和技术发展规划纲要 (2006~2020 年)》已将"量子调控研究"列为基础研究的四项重大研究计划之一. 2017 年 5 月印发的"十三五"国家基础研究专项规划中, "量子通信与量子计算机"和"量子调控与量子信息"分别为国家组织实施的重大科技项目和加强战略性前瞻性重大科学问题之首要任务. 欧盟委员会也于 2016 年 4 月提出"量子宣言", 在 2018 年启动十亿欧元量子技术项目, 希望借此促进包括安全的通信网络和通用量子计算机等在内的多项量子技术的发展. 对量子操控的研究可以使人们实现对物理学基本原理的掌控, 并最终实现新型的人工量子器件或物质.

1.2.2 量子相干操控概念

本小节将简要介绍什么是量子相干操控 (quantum coherent manipulation). 激光冷却和陷俘原子的技术为人类操控微观粒子, 尤其是电中性的原子, 打开了一扇全新的大门[40]. 对一个量子系统而言, 波函数包含了系统的全部信息. 一旦知道了系统的波函数, 就可以知道系统的整个物理状态, 如位置、动量、角动量以及能量等. 从波函数的层次上随心所欲地操控量子系统, 是人们梦寐以求的目标. 量子相干操控正是为此而发展起来的一个重要研究课题. 所谓量子相干操控, 是指对处于已知初始状态, 如已知初始波函数、初始粒子数分布等的量子系统, 利用特定的外场, 如激光场、外加磁场以及射频场等来操控系统的动力学过程或行为, 以实现人们所需要的目标状态. 它是根据量子力学的基本规律主动地去制造、改变和控制体系的波函数或量子态.

1.2.3 超冷原子系统中量子操控的方法

对于超冷原子系统而言, 其操控原理和方法主要有以下几种[41].

(1) 利用空间物质结构–原子的相互作用. 在这种方法中, 当原子的 de Broglie 波长大于空间物质结构的特征尺度时, 原子的波动性就成为主要因素. 特别是对玻色–爱因斯坦凝聚而言, 由于温度很低, de Broglie 波长大于原子间的平均距离, 原子间会相互交叠, 高度相干, 如图 1.5 所示. 这样超冷原子系统就成了近年来量子相干操控的首选, 这类方法中典型的有物质"光栅"对原子的衍射, 以及原子全息术等.

(2) 利用静电场、静磁场–原子的相互作用. 具有感生电矩或磁矩的原子在非均匀静电场或静磁场中会受到力的作用, 由一个势场来控制, 通过设计一个合适空间分布的静电场或静磁场, 就能利用这种相互作用对原子进行控制. 这类方法中典型的有原子磁阱、原子的磁导引、原子芯片等. 这里特别指出, 磁场调节的 Feshbach 共振、Efimov 共振等 (2.3 节将会专门介绍), 可用来研究量子系统的隧穿效应、超冷原子到分子的转化等 (本章后几节将专门介绍).

(3) 利用光场–原子的相互作用. 光场是高频振荡的电磁场, 原子在场中的受力主要有两类: 当光的频率与原子的本征频率远离共振时, 原子感受到一个光势, 受到的力主要是偶极梯度力; 当近共振时, 原子受到的力主要是自发辐射力, 因而可设

1.2 量子相干操控

计出不同分布的光场来控制原子的运动. 这类方法主要包括原子冷却、光学黏胶、原子的空心光纤导引、原子被光场的衍射、光阱、衰逝场原子反射镜、空心光束导引、原子喷泉、原子分束、原子制版等典型的实际应用. 另外, 如光缔合技术、受激拉曼绝热通道技术等, 可用于形成超冷分子.

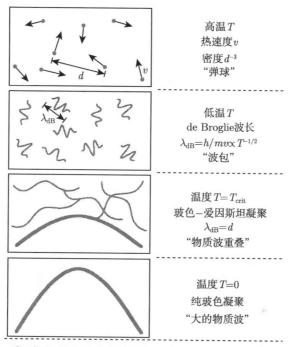

图 1.5 玻色–爱因斯坦凝聚形成条件及其过程中 de Broglie 波长变化示意图 [42]

(4) 以上几种方法相结合, 如 Feshbach 共振和光缔合相结合技术 [43-48]. 这些组合方式集各种方法的优点于一身, 其应用性更强, 如 Feshbach 共振援助的光缔合技术可以实现超冷原子到稳定分子的转化 [49,50] (2.3 节将专门介绍).

这些发展促进了原子光学这一新的研究领域的形成和进步. 例如, 原子反射镜 [51], 它实际上是让原子在一个保守力场下完成被减速到零再加速的过程; 原子透镜 [52], 它利用磁场或光场对原子施加的力来操控原子的运动完成聚焦, 产生的高强度原子源可用来完成沉积和刻蚀等工艺, 在微细加工和纳米技术方面有广泛的应用. 而原子导引实际上是用力场将要输运的原子悬浮起来, 再利用原子的初速度、运动黏团的驱动力、原子的反射 [53]、力场的移动来使原子转移.

量子操控技术的巨大进步使得人类可以对光子、原子等微观粒子进行精确调控, 从而能够以一种全新的方式利用量子规律. 特别是在量子信息科学中, 如量子通信、量子计算以及量子精密测量等方面将大显身手, 在确保信息安全、提高运算速度、提升测量精度等许多重要方面突破经典信息技术的瓶颈, 为社会紧急发展面临的若干重大问题提供革命性的解决途径.

与此同时, 激光技术的发展也将进一步深刻地影响对原子系统的相干操控. 目前, 激光技术不断取得突破, 实现超高频率、超高功率、超大能量密度、超短脉冲等. 例如, 超高频的"自由电子激光"就是要建立比现有同步辐射光源强数十亿倍的 X 波段的相干光源, 其产生的电场强度很强, 远超过原子内的电场强度, 从而可以把分子、原子击成碎片. 利用激光形成的超大能量密度可以用来模拟早期宇宙的演化和恒星内部的物理过程, 为可控核聚变过程"点火". 在超短脉冲方面, 目前激光脉冲宽度可以短到一个飞秒 (10^{-15}s) 量级, 甚至几十个阿秒 (10^{-18}s) 量级, 利用如此短的激光脉冲就可以直接观察分子和原子内部的电子过程, 用来测量单个分子的动态瞬时结构, 甚至控制化学反应的路径等[54]. 这些激光用在超冷原子系统中, 来控制超冷原子分子, 势必会为超冷原子系统的操控增添新的活力.

1.3 超冷原子系统量子操控的研究意义

得益于超冷原子气体的诸多优异性质, 如量子力学波动性、宏观量子相干性以及人工精确可调控性等, 冷原子物理从产生就和量子操控紧密地结合起来. 特别是随着量子信息和一些相关实验技术的发展, 基于冷原子体系的物理研究已完全突破了对现有理论的验证阶段, 正朝着全面发展新的理论和概念, 探索新的自然规律和新的应用方向发展.

超冷原子物理, 特别是原子玻色–爱因斯坦凝聚体的诞生, 不仅为研究原子物理、量子理论和多体系统开辟了新窗口, 还发现了许多重要的实验现象, 如物质波放大器[55]、物质波孤子[56]、涡旋[57] 以及光晶格中的量子相变[58] 等.

在基础科学领域, 冷原子量子操控研究是研究冷碰撞和超冷碰撞[59]、原子光学、冷原子操控[40] 等方面的绝佳工具, 为研究少体和多体物理、量子信息等领域提供了极好的环境, 也为研究"负温度"物质铺平了道路. 2013 年 1 月, 德国物理学家 Braun 及其合作者利用超冷原子气体的特殊性质, 在超冷钾原子量子气体中首次制造出一种负热力学温度的量子气体[60,61], 如图 1.6 所示, 图中第一行表示热力学温度 T, 第二行表示 $-\beta = -1/k_\text{B}T$, 第三行表示熵 S 随能量从最小 (E_min) 到最大 (E_max) 的变化情况. 小图表示弱相互作用下, 单粒子态在正、无穷和负温度下的布居数分布[60]. 在热学中, 温度的定义为 $1/T = \partial S/\partial E$, 当 $T = 0$ 时, 系统处在最低能态 E_min, 粒子处在基态上, 仅基态上有布居数分布, 系统的熵为零. 随着温度的增加, 系统的能量和熵增加. 当 $T = \infty$ 时, 系统能量为零, 各态上布居数相等, 熵达到最大值 S_max (温度在最大熵时是不连续的, 从正无穷跳变到负无穷), 此时 $T = +\infty$ 和 $T = -\infty$ 等价. 系统能量继续增加, 处于高能级上的粒子数超过处于低能级上的粒子数越来越多时, 熵会下降, 从而导致温度变负, 相当于温度从

$T = -\infty$ 继续升高, 绝对值继续减小. 到 $T = -0$ 时, 能量到最大值, 粒子都处在高能级上, 熵变为零. 科学家称这一成果为"实验的绝技", 为将来实现负温度物质以及新型量子器件等奠定了基础. 超冷原子的玻色–爱因斯坦凝聚体通过和凝聚态物理等传统领域相结合, 不断地促生新的交叉领域前沿, 解决复杂凝聚态物理中的疑难问题. 通过控制超冷原子间的相互作用, 还能制备振动–转动基态的超冷分子, 为研究化学反应微观机制、长程相互作用、量子磁性等提供了可靠和精确的实验平台.

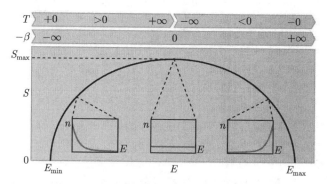

图 1.6 光晶格中的负热力学温度 [60]

在应用方面, 中性冷原子及其量子操控在精密测量领域有着无可替代的优势, 超冷原子技术是实现当今世界上最精确的原子钟的重要技术手段 [62], 是全球卫星定位导航系统中的关键性核心技术. 我国北斗三号卫星导航系统中就采用新一代高精度铷原子钟来提供时间基准, 大大提高了定位精度和稳定度. 截至 2018 年 12 月, 我国已成功发射四十三颗北斗导航卫星. 2016 年 8 月 16 日, 人类历史上第一颗用于量子通信研究的"墨子号"量子科学实验卫星在我国酒泉卫星发射中心发射升空. 该卫星是世界上第一颗从事空间尺度量子科学实验的卫星. 其主要任务有: 配合多个地面站, 在国际上率先实现星地高速量子密钥分发、星地双向量子纠缠分发及空间尺度量子非定域性检验 (量子力学基本原理的检验)、地星量子隐形传态, 以及探索广域量子密钥组网等实验. 2016 年 9 月 15 日, 我国第一个真正意义上的空间实验室 —— 天宫二号发射成功. 在该空间实验室上安排了涵盖基础物理、空间天文、微重力科学、空间生命科学和地球科学观测及应用等 14 项前沿科学研究与应用任务. 其中, 搭载的空间冷原子钟, 如图 1.7 所示, 其频率稳定度达到 10^{-16} 量级 (即每 3000 万年误差小于 1s), 成为国际上首次在轨运行的空间冷原子钟, 不仅为空间高精度时频系统、空间冷原子物理、空间冷原子干涉仪、空间冷原子陀螺仪等各种量子敏感器奠定技术基础, 而且在全球卫星导航定位系统、深空探测、广义相对论验证、引力波测量、地球重力场测量、基本物理常数测量等一系列重大技术和科学发展方面做出重要贡献 [63]. 图 1.8 和图 1.9 分别展示了时间装置演变图以及空间冷原子钟的典型应用. 另外, 超冷原子的玻色-爱因斯坦凝聚体正在成为量子计算机和费曼量子模拟器的重要实验平台, 在量子计算中发挥着至关重要的作用. 近几年来, 国际上展开了一场关于量子计算机研究的角逐, 许多新的成果不断涌现, 令人惊喜. 2008 年, Daley 等 [64] 基于碱土金属 Sr 发展了一种光晶格量子计算机的理论构架, 在研究 Sr 晶格光原子钟的过程中, Ye 研究小组已经探索了 Sr 原子的许多方面, 包括冷却、电子转移以及光晶格等, 为量子信息过程中的一些关键问题, 如量子比特的存储、访问以及转移等提出了一些解决方案, 见图 1.10[65]. 计算前, 单个 Sr 原子作为量子比特 (qubits) 被安置在存储晶格 (storage lattice) (光阱) 中, 精密钟激光将用于移动量子比特到传输晶格 (transport lattice), 它们能被投递到原子相互作用的特殊位置, 完成计算后, 激光再次把量子比特返回到存储晶格中 [65]. 2017 年 5 月 3 日, 我国宣布在量子计算机研究方面取得突破性进展, 世界上第一台超越早期经典计算机的量子计算机 (量子模拟机) 诞生, 为最终实现超越经典计算能力的量子计算奠定了基础. 最近更是好消息不断, 许多国家或企业陆续宣布有新突破, 如 IBM、谷歌等, 一场实现量子霸权的较量正在悄然兴起. 另外, 基于超冷原子的原子激射器 [18,66]、原子干涉仪 [67]、冷原子陀螺仪、量子加速计以及基于原子量子效应的高精度惯性导航技术 —— 量子导航和量子计算机等方面的研究也将进一步在重力加速度测量等基础物理领域 [68], 以及国防、航空等方面发挥重要作用. 该领域的研究成果不仅为探索极端条件下的微观世界开启了大门, 也为精密测量、纳米技术等带来了革命性的变化, 其关系到国家安全和未来高新科学技术的发展. 总之, 对超冷原子物理及其量子操控的研究, 不仅有十分重要的科学研究价值, 而且在人类生活、技术进步、国民经济可持续发展战略以及国家安全等领域均占有非常重要的地位. 因此, 超冷原子物理的量子操控成为原子、分子和光物理等领域近年来最重要、最活跃、发展最为迅速、成果最为辉煌的前沿课题之一.

图 1.7 超高精度空间冷原子钟 [63]

1.3 超冷原子系统量子操控的研究意义

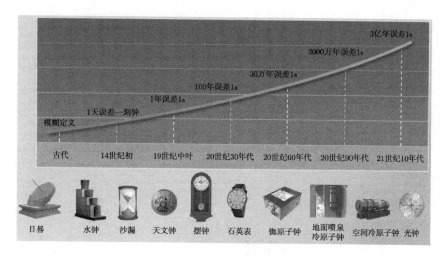

图 1.8 时间装置的演变图 [63]

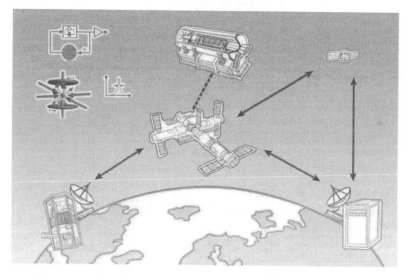

图 1.9 空间冷原子钟的典型应用 [63]

尽管传统的量子相干操控技术已经超过运用于超冷原子系统的许多方面,但是,随着人们研究的不断深入,发现由于超冷原子系统的特殊性,如玻色–爱因斯坦凝聚中粒子间相互作用等,这种粒子间的非线性相互作用使得一系列原来的传统方法随之失效,同时会出现许多新奇现象.因此,通过发展或创新量子操控技术探究超冷原子世界,将势在必行.第 2 章将介绍本书所涉及的超冷原子隧穿动力学中几个重要方面的研究现状和基本方法.

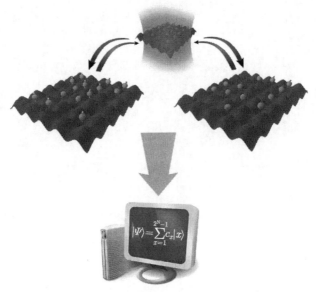

图 1.10　基于 Sr 原子的光晶格量子计算机[65]

参 考 文 献

[1] ONNES H K. The resistance of pure mercury at helium temperatures[J]. Commun. Phys. Lab. Univ. Leiden, 1911, 12: 120.

[2] KAPITZA P. Viscosity of liquid helium below the λ-point[J]. Nature, 1938, 141: 74.

[3] OSHEROFF D D, RICHARDSON R C, LEE D M. Evidence for a new phase of solid He^3[J]. Phys. Rev. Lett., 1972, 28: 885-888.

[4] CHU S. Nobel Lecture: The manipulation of neutral particles[J]. Rev. Mod. Phys., 1998, 70: 685-706.

[5] COHEN-TANNOUDJI C N. Nobel Lecture: Manipulating atoms with photons[J]. Rev. Mod. Phys., 1998, 70: 707-719.

[6] PHILLIPS W D. Nobel Lecture: Laser cooling and trapping of neutral atoms[J]. Rev. Mod. Phys., 1998, 70: 721-741.

[7] ANDERSON M H, ENSHER J R, MATTHEWS M R, et al. Observation of Bose-Einstein condensation in a dilute atomic vapor[J]. Science, 1995, 269(5221): 198-201.

[8] ESSLINGER T, BLOCH I, HÄNSCH T W. Bose-Einstein condensation in a quadrupole-Ioffe-configuration trap[J]. Phys. Rev. A, 1998, 58: R2664-R2667.

[9] BRADLEY C C, SACKETT C A, TOLLETT J J, et al. Evidence of Bose-Einstein condensation in an atomic gas with attractive interactions[J]. Phys. Rev. Lett., 1995, 75: 1687-1690.

[10] DAVIS K B, MEWES M O, ANDREWS M R, et al. Bose-Einstein condensation in a gas of sodium atoms[J]. Phys. Rev. Lett., 1995, 75: 3969-3973.

参考文献

[11] WELD D M, MEDLEY P, MIYAKE H, et al. Spin gradient thermometry for ultracold atoms in optical lattices[J]. Phys. Rev. Lett., 2009, 103: 245301.

[12] 刘杰. 玻色-爱因斯坦凝聚体动力学 —— 非线性隧穿、相干及不稳定性 [M]. 北京: 科学出版社, 2009.

[13] 豆福全. 超冷原子系统中非线性隧穿及原子-分子转化的量子操控 [D]. 北京: 北京理工大学, 2013.

[14] OJA A, LOUNASMAA O. Nuclear magnetic ordering in simple metals at positive and negative nanokelvin temperatures[J]. Rev. Mod. Phys., 1997, 69(1): 1.

[15] TUORINIEMI J. Physics at its coolest[J]. Nature Phys., 2016, 12(1): 11-14.

[16] JILA. Cold and ultracold[OL]. [2013-03-13]. http://jila.colorado.edu/research/atomic-and-molecular-physics/cold-molecules.

[17] BOSE S. Plancks Gesetz und Lichtquantenhypothese[J]. Zeitschrift für Physik a Hadrons and Nuclei, 1924, 26: 178-181.

[18] KETTERLE W. Nobel lecture: When atoms behave as waves: Bose-Einstein condensation and the atom laser[J]. Rev. Mod. Phys., 2002, 74: 1131-1151.

[19] FRIED D G, KILLIAN T C, WILLMANN L, et al. Bose-Einstein condensation of atomic hydrogen[J]. Phys. Rev. Lett., 1998, 81: 3811-3814.

[20] CORNISH S L, CLAUSSEN N R, ROBERTS J L, et al. Stable ^{85}Rb Bose-Einstein condensates with widely tunable interactions[J]. Phys. Rev. Lett., 2000, 85: 1795-1798.

[21] ROBERT A, SIRJEAN O, BROWAEYS A, et al. A Bose-Einstein condensate of metastable atoms[J]. Science, 2001, 292(5516): 461-464.

[22] PEREIRA DOS SANTOS F, LÉONARD J, WANG J, et al. Bose-Einstein condensation of metastable helium[J]. Phys. Rev. Lett., 2001, 86: 3459-3462.

[23] MODUGNO G, FERRARI G, ROATI G, et al. Bose-Einstein condensation of potassium atoms by sympathetic cooling[J]. Science, 2001, 294(5545): 1320-1322.

[24] WEBER T, HERBIG J, MARK M, et al. Bose-Einstein condensation of cesium[J]. Science, 2003, 299(5604): 232-235.

[25] TAKASU Y, MAKI K, KOMORI K, et al. Spin-singlet Bose-Einstein condensation of two-electron atoms[J]. Phys. Rev. Lett., 2003, 91: 040404.

[26] YAMAZAKI R, TAIE S, SUGAWA S, et al. Submicron spatial modulation of an interatomic interaction in a Bose-Einstein condensate[J]. Phys. Rev. Lett., 2010, 105: 050405.

[27] GRIESMAIER A, WERNER J, HENSLER S, et al. Bose-Einstein condensation of chromium[J]. Phys. Rev. Lett., 2005, 94: 160401.

[28] ROATI G, ZACCANTI M, D'ERRICO C, et al. ^{39}K Bose-Einstein condensate with tunable interactions[J]. Phys. Rev. Lett., 2007, 99: 010403.

[29] FUKUHARA T, SUGAWA S, TAKAHASHI Y. Bose-Einstein condensation of an ytterbium isotope[J]. Phys. Rev. A, 2007, 76: 051604.

[30] KRAFT S, VOGT F, APPEL O, et al. Bose-Einstein condensation of alkaline earth atoms: ^{40}Ca[J]. Phys. Rev. Lett., 2009, 103: 130401.

[31] STELLMER S, TEY M K, HUANG B, et al. Bose-Einstein condensation of strontium[J]. Phys. Rev. Lett., 2009, 103: 200401.

[32] DE ESCOBAR Y N M, MICKELSON P G, YAN M, et al. Bose-Einstein condensation of ^{84}Sr[J]. Phys. Rev. Lett., 2009, 103: 200402.

[33] MICKELSON P G, MARTINEZ DE ESCOBAR Y N, YAN M, et al. Bose-Einstein condensation of ^{88}Sr through sympathetic cooling with ^{87}Sr[J]. Phys. Rev. A, 2010, 81: 051601.

[34] STELLMER S, TEY M K, GRIMM R, et al. Bose-Einstein condensation of ^{86}Sr[J]. Phys. Rev. A, 2010, 82: 041602.

[35] AIKAWA K, FRISCH A, MARK M, et al. Bose-Einstein condensation of erbium[J]. Phys. Rev. Lett., 2012, 108: 210401.

[36] CORNELL E A, WIEMAN C E. Nobel Lecture: Bose-Einstein condensation in a dilute gas, the first 70 years and some recent experiments[J]. Rev. Mod. Phys., 2002, 74: 875-893.

[37] PETHICK C J, SMITH H. Bose-Einstein Condensation in Dilues Gases[M]. UK: Cambridge University Press, 2008.

[38] 詹明生. 冷原子物理 [J]. 中国科学院院刊, 2002, 6: 407-412.

[39] THE NOBEL PRIZE. The Nobel prize[OL]. [2013-03-18]. http://www.nobelprize.org/.

[40] CHU S. Laser manipulation of atoms and particles[J]. Science, 1991, 253(5022): 861-866.

[41] 李代军, 刘夏姬, 胡正峰, 等. 超冷原子的操控、导引及其应用 [J]. 量子光学学报, 2000, 6(3): 122.

[42] KETTERLE W. What is BEC?[OL]. [2013-06-18]. http://www.rle.mit.edu/cua-pub/ketterle-group/intro/whatbec/whtisbec.html.

[43] THEIS M, THALHAMMER G, WINKLER K, et al. Tuning the scattering length with an optically induced Feshbach resonance[J]. Phys. Rev. Lett., 2004, 93: 123001.

[44] BAUER D M, LETTNER M, VO C, et al. Control of a magnetic Feshbach resonance with laser light[J]. Nature Phys., 2009, 5: 339.

[45] FU Z, WANG P, HUANG L, et al. Optical control of a magnetic Feshbach resonance in an ultracold Fermi gas[J]. Phys. Rev. A, 2013, 88: 041601.

[46] YAN M, DESALVO B J, RAMACHANDHRAN B, et al. Controlling Condensate collapse and expansion with an optical Feshbach resonance[J]. Phys. Rev. Lett., 2013, 110: 123201.

[47] ZHANG P, NAIDON P, UEDA M. Independent control of scattering lengths in multicomponent quantum gases[J]. Phys. Rev. Lett., 2009, 103: 133202.

[48] ZHANG Y C, LIU W M, HU H. Tuning a magnetic Feshbach resonance with spatially modulated laser light[J]. Phys. Rev. A, 2014, 90: 052722.

[49] KOCH C P, MASNOU-SEEUWS F M C, KOSLOFF R. Creating ground state molecules with optical Feshbach resonances in tight traps[J]. Phys. Rev. Lett., 2005, 94: 193001.

[50] TAIE S, WATANABE S, ICHINOSE T, et al. Feshbach-resonance-enhanced coherent atom-molecule conversion with ultra-narrow photoassociation resonance[J]. Phys. Rev. Lett., 2016, 116: 043202.

[51] BALYKIN V I, LETOKHOV V S, OVCHINNIKOV Y B, et al. Quantum-state-selective mirror reflection of atoms by laser light[J]. Phys. Rev. Lett., 1988, 60: 2137-2140.

[52] KAENDERS W, LISON F, RICHTER A, et al. Imaging with an atomic beam[J]. Nature, 1995, 375: 214.

参考文献

[53] RENN M J, MONTGOMERY D, VDOVIN O, et al. Laser-guided atoms in hollow-core optical fibers[J]. Phys. Rev. Lett., 1995, 75: 3253-3256.

[54] 于渌. 量子调控和相关研究的若干进展与展望[J]. 科学新闻, 2007, (7): 4-5.

[55] INOUYE S, PFAU T, GUPTA S, et al. Phase-coherent amplification of atomic matter waves[J]. Nature, 1999, 402: 641-644.

[56] DENSCHLAG J, SIMSARIAN J, FEDER D, et al. Generating solitons by phase engineering of a Bose-Einstein condensate[J]. Science, 2000, 287(5450): 97-101.

[57] MATTHEWS M R, ANDERSON B P, HALJAN P C, et al. Vortices in a Bose-Einstein condensate[J]. Phys. Rev. Lett., 1999, 83: 2498-2501.

[58] GREINER M, MANDEL O, ESSLINGER T, et al. Quantum phase transition from a superfluid to a Mott insulator in a gas of ultracold atoms[J]. Nature, 2002, 415(6867): 39-44.

[59] WEINER J, BAGNATO V S, ZILIO S, et al. Experiments and theory in cold and ultracold collisions[J]. Rev. Mod. Phys., 1999, 71: 1-85.

[60] BRAUN S, RONZHEIMER J P, SCHREIBER M, et al. Negative absolute temperature for motional degrees of freedom[J]. Science, 2013, 339(6115): 52-55.

[61] CARR L D. Negative temperatures?[J]. Science, 2013, 339(6115): 42-43.

[62] DIDDAMS S A, BERGQUIST J C, JEFFERTS S R, et al. Standards of Time and frequency at the outset of the 21st century[J]. Science, 2004, 306(5700): 1318-1324.

[63] 乔勇军, 刘伍明. 天宫二号里那块优雅的"手表"——空间冷原子钟[J]. 自然杂志, 2017, 39(1): 54-61.

[64] DALEY A J, BOYD M M, YE J, et al. Quantum computing with alkaline-earth-metal atoms[J]. Phys. Rev. Lett., 2008, 101: 170504.

[65] YE J, REY A M, KUEBLER G. Artist's concept of a Sr-lattice-based quantum computer[OL]. [2018-01-10]. https://jila.colorado.edu/research/quantum-information/quantum-computing.

[66] MEWES M O, ANDREWS M R, KURN D M, et al. Output coupler for Bose-Einstein condensed atoms[J]. Phys. Rev. Lett., 1997, 78: 582-585.

[67] JAVANAINEN J, YOO S M. Quantum phase of a Bose-Einstein condensate with an arbitrary number of atoms[J]. Phys. Rev. Lett., 1996, 76: 161-164.

[68] PETERS A, CHUNG K Y, CHU S. Measurement of gravitational acceleration by dropping atoms[J]. Nature, 1999, 400: 849-852.

第 2 章 隧穿动力学：量子操控重要方向

第 1 章介绍了超冷原子系统量子隧穿相关的研究背景、研究意义及其相关概念. 本章主要对本书涉及"超冷原子量子操控"的三个方面进行简要综述, 包括超冷原子系统中的非线性隧穿动力学、高保真度量子操控、超冷原子–分子转化.

2.1 非线性隧穿动力学

2.1.1 基本模型

隧穿效应是一种基本的量子现象, 在化学反应系统、核物理系统、自旋体系、超晶格[1]、超导器件[2], 特别是超冷原子系统中有许多重要的应用[3]. 为便于区分各种不同类型的隧穿行为, 首先写出有限模近似下描述超冷原子系统的离散形式的 Gross-Pitaevskii 方程[4,5]：

$$i\frac{\partial}{\partial t}\begin{pmatrix} a_1 \\ a_2 \\ \vdots \\ a_N \end{pmatrix} = H(t)\begin{pmatrix} a_1 \\ a_2 \\ \vdots \\ a_N \end{pmatrix}, \tag{2.1}$$

哈密顿量为

$$H(t) = \begin{pmatrix} \gamma_1 + c|a_1|^2 & v_{12} & \cdots & v_{1N} \\ v_{21} & \gamma_2 + c|a_2|^2 & \cdots & v_{2N} \\ \vdots & \vdots & & \vdots \\ v_{N1} & v_{N2} & \cdots & \gamma_N + c|a_N|^2 \end{pmatrix}, \tag{2.2}$$

其中, a_j 是第 j 个态的概率幅, 满足归一化条件: $\sum |a_j(t)|^2 = 1$; N 是总粒子数. 哈密顿量中矩阵元分别是

$$\begin{cases} \gamma_j = \int \frac{\hbar^2}{2m}|\nabla \varphi_j(\boldsymbol{r})|^2 \mathrm{d}\boldsymbol{r} + \int V_{\mathrm{ext}}(\boldsymbol{r})|\varphi_j(\boldsymbol{r})|^2 \mathrm{d}\boldsymbol{r}, \\ v_{jk} = \int \frac{\hbar^2}{2m}\nabla \varphi_j^*(\boldsymbol{r})\varphi_k(\boldsymbol{r}) \mathrm{d}\boldsymbol{r} + \int V_{\mathrm{ext}}(\boldsymbol{r})\varphi_j^*(\boldsymbol{r})\varphi_k(\boldsymbol{r}) \mathrm{d}\boldsymbol{r} \quad (j \neq k), \end{cases} \tag{2.3}$$

其中, γ_j 表示第 j 个量子态的本征能; v_{jk} 表示第 j 个量子态和第 k 个量子态的耦合强度; $\varphi_j(\boldsymbol{r})$ 表示第 j 个量子态或第 j 个阱中基态波函数; m 为原子的质量; V_{ext}

表示囚禁势. 参数 c 至关重要, 表示原子之间的相互作用:

$$c = gN = \frac{4\pi\hbar^2 a}{m}N, \quad (2.4)$$

其中, a 表示 s 波散射长度. 该项的参与使得系统表现出更加复杂而丰富的物理性质. 本书后几章会讨论具有这种非线性相互作用的超冷原子系统的一系列有趣行为. 这里的非线性隧穿, 也就是系统中包含了诸如此项 (不仅仅只含粒子间相互作用项) 的非线性项时系统的隧穿动力学.

最典型实用的情形是两能级系统, 它是一个"简约而不简单 (simplest non-simple)"的模型, 能代表许多具体的物理系统. 此时, Gross-Pitaevskii 方程 (2.1) 就变为

$$i\frac{\partial}{\partial t}\begin{pmatrix} a_1 \\ a_2 \end{pmatrix} = \begin{pmatrix} \gamma_1 + c|a_1|^2 & v_{12} \\ v_{21} & \gamma_2 + c|a_2|^2 \end{pmatrix}\begin{pmatrix} a_1 \\ a_2 \end{pmatrix}. \quad (2.5)$$

若取 $v_{12} = v_{21} = v/2$, 主对角项同时减去常数项 $(\gamma_1 + \gamma_2 + c)/2$, 并令 $\gamma_1 - \gamma_2 = \gamma$, 则方程 (2.5) 变为

$$i\frac{\partial}{\partial t}\begin{pmatrix} a_1 \\ a_2 \end{pmatrix} = \begin{pmatrix} \frac{\gamma}{2} + \frac{c}{2}(|a_1|^2 - |a_2|^2) & \frac{v}{2} \\ \frac{v}{2} & -\frac{\gamma}{2} - \frac{c}{2}(|a_1|^2 - |a_2|^2) \end{pmatrix}\begin{pmatrix} a_1 \\ a_2 \end{pmatrix}, \quad (2.6)$$

其中, 用到了归一化条件 $|a_1|^2 + |a_2|^2 = 1$. 方程 (2.6) 就是本书中常用的二能级模型.

对于量子系统而言, 本征态具有至关重要的意义. 于是, 定义本征态为

$$\begin{pmatrix} \frac{\gamma}{2} + \frac{c}{2}(|a_1|^2 - |a_2|^2) & \frac{v}{2} \\ \frac{v}{2} & -\frac{\gamma}{2} - \frac{c}{2}(|a_1|^2 - |a_2|^2) \end{pmatrix}\begin{pmatrix} a_1 \\ a_2 \end{pmatrix} = \mu\begin{pmatrix} a_1 \\ a_2 \end{pmatrix}, \quad (2.7)$$

其中, μ 是本征值, 也称为化学势, 即往系统里加入一个粒子所需要的能量. 由此定义可以获得系统的能级结构.

2.1.2 几种常见的隧穿模型

下面, 首先讨论不包含粒子间相互作用项, 即非线性项时, 由于 γ_j 和 v_{jk} (或者 γ 和 v) 取不同形式所对应的几种常见物理模型.

(1) 拉比 (Rabi) 振荡 (或 Josephson 振荡). γ 和 v 均为常数, 此模型早在 20 世纪 30 年代期间由 Rabi 等研究, Rabi 发表了一系列和此有关的文章.

(2) 复合脉冲模型 (composite pulse model). v 为常数，γ 除始末值为固定的常数外，其他时候均为零.

(3) Landau-Zener 隧穿. v 为常数，γ 随时间线性变化. 该模型早在 1932 年由 4 位物理学家: Landau[6]、Zener[7]、Stückelberg[8] 和 Majorana[9] 在理论上独立发现的, 因此合理的叫法应该是 Landau-Zener-Stückelberg-Majorana 模型[10]. Landau 和 Zener 在研究分子碰撞问题中引出该模型, Landau 利用扰动法主要研究非简并系统中近绝热极限下的跃迁概率, 而 Zener 主要研究分子的两个基本状态: 激化态和非激化态之间的跃迁. 假定系统初始态 $a_1(-\infty) = 1, a_2(-\infty) = 0$, 得到了单通过 (single passage) 情况下的隧穿表达式 (绝热基下):

$$P_{\mathrm{LZ}} = \exp(-\varsigma), \tag{2.8}$$

这就是著名的 Landau-Zener 公式. 其中 $\varsigma = \dfrac{\pi v^2}{2\hbar\dot\gamma}$，($\dot{}$ 表示对时间的导数) 也常常表示 Landau-Zener 参数. Stückelberg 考虑的是具有动能和势能项的哈密顿量, 采用 WKB (Wentzel-Kramers-Brillouin) 近似, 得到了双通道 (double passage) 时包含 Stückelberg 相的概率公式, 对于单通过情形, 对应的跃迁概率和式 (2.8) 一致. 而 Majorana 主要研究在随时间变化的磁场中, 电子的自旋翻转问题, 其哈密顿量为

$$H(t) = g\dot{B}_z t \hat{s}_z + gB_x \hat{s}_x, \tag{2.9}$$

其中, g 是回旋磁比率; B_z, B_x 分别表示 z, x 方向的磁场; $\hat{s}_x$ 和 $\hat{s}_z$ 是自旋算符; $\dot{}$ 表示对时间的导数. 可得自旋翻转的概率公式为

$$P_{\mathrm{M}} = \exp\left(-\dfrac{\pi g B_x^2}{\hbar \dot{B}_z}\right), \tag{2.10}$$

可以看到, 当 $\varsigma = \dfrac{\pi g B_x^2}{\hbar \dot{B}_z}$, 即 $v = \sqrt{2g}B_x, \gamma = B_z$ 时, 式 (2.10) 和式 (2.8) 也一致. 遗憾的是, Majorana 的名字从没有在式 (2.8) 中被提到, 而他的方法却别出心裁, 是 Landau 方法、Zener 方法、Stückelberg 方法的补充. 因此, Giacomo 和 Nikitin[10] 专门写了一篇文章说明此事, 并在文章结尾写到: "我们希望这篇文章可以纠正一些历史上对 Majorana 在非绝热跃迁方面贡献的忽视, 遗憾的是, 包括我们自己文章在内的当前文献中并未给予认可 (We hope that this note will partly rectify historical neglect of the Majorana's contribution to the theory of nonadiabatic transitions, which, regrettably, is not properly acknowledged in current reference lists, including those by one of the authors of this paper)".

(4) Rosen-Zener 跃迁. γ 为常数, $v = \Omega_0 \mathrm{sech}(t/T)$. 该模型起源于 1932 年 Rosen 和 Zener[11] 在研究双 Stern-Gerlach 实验中, 原子在旋转磁场中的自旋反转问题.

(5) Demkov-Kunike 模型, 分为第一类和第二类 Demkov-Kunike 模型[12]. $v = \Omega_0 \mathrm{sech}(t/T)$, $\gamma = \gamma_0 \tanh(t/T)$ (该形式也称为 Allen-Eberly 模型[13]) 或 $\gamma = a + \gamma_0 \tanh(t/T)$ (第一类); 如果 v 为常数, $\gamma = \gamma_0 \tanh(t/T)$, 则称为第二类 Demkov-Kunike 模型. 该模型最早由 Demkov 和 Kunike[14] 于 1969 年提出, 后来由许多人所研究, 如 Hioe 和 Carroll[15] 以及 Kyoseva 和 Vitanov[16] 等. 另外, 在第一类 Demkov-Kunike 模型中, 如果参数 $a = \gamma_0$, 则称为 Bambini-Berman 模型[17].

(6) Nikitin 指数模型, 也分为第一类和第二类. 若 v 为常数, $\gamma = \gamma_0 \exp(-at)$ 为第一类 Nikitin 指数模型; 若 $v = \Omega_0 \exp(-at)$, $\gamma = \gamma_0 \exp(-at)$, 则为第二类 Nikitin 指数模型. 该模型最早于 1962 年由 Nikitin[18] 提出, 后来陆续有些扩展[19]. 然而, 到目前为止, 该模型不能精确求解, 只能获得近似解.

(7) t^n 模型[20]. v 为常数, $\gamma = \gamma_0(t/T)^n (n = 1, 2, 3, \cdots)$.

(8) 相跳模型 (phase jump model)[21]. $\gamma = a + \gamma_0 \tanh(t/T)$, 而 $v = \Omega_0 \mathrm{sech}(t/T)$ $(t < 0), v = \exp(\mathrm{i}\varphi) \Omega_0 \mathrm{sech}(t/T) (t \geqslant 0)$.

(9) 受激拉曼绝热通道 (stimulated Raman adiabatic passage, STIRAP)[22]. γ_1、γ_2、γ_3 为常数, $v_{12} = v_{21} = \Omega_0 \mathrm{sech}[(t-t_1)/T_1]$, $v_{23} = v_{32} = \Omega_2 \mathrm{sech}[(t-t_2)/T_2]$.

(10) 其他模型. 例如, 高斯脉冲模型[23]、Roland-Cerf 模型[24], 以及以上模型的各种拓展.

实际上, 可以看到针对一个量子系统 (2.1), 对应的模型可以多种多样, 但是不外乎这样四种情况: 其一, v、γ 均为常数; 其二, v 为常数, γ 随时间变化; 其三, γ 为常数, v 随时间变化; 其四, v、γ 均随时间变化. 原则上, v 和 γ 可以是随时间变化的任意函数, 然而, 在真实物理系统中, 有些变化形式没有实际物理意义. 另外要获得精确解析解, 对 v、γ 的具体形式要求比较苛刻. 当不考虑粒子间相互作用时, 发现以上模型大部分有精确解析解.

2.1.3 非线性方面的扩展

以上各种模型对应于各种物理系统, 在很多领域都有着广泛的应用, 如分子纳米磁子[25,26]、分子碰撞[27]、量子点阵[28]、里德堡原子[29]、光晶格[30]、准粒子激发[31]、超晶格[1]、Josephson 结[32]、光学定向耦合器[33]、耦合波导管[34], 特别是超冷原子和分子系统等[35,36], 并在耦合持续效应系统[37]、多态系统[38,39]、噪声和耗散系统[40-42]、多体量子系统[43] 等有所发展. 然而, 在真实的物理系统中, 会出现包含粒子间相互作用等非线性项. 特别是在玻色–爱因斯坦凝聚系统中, 通过对原子间相互作用作平均场近似, 很自然会得到此项. 因此, 近年来有关模型在非线性方面的拓展也格外引人注目.

最引人注目的工作之一是 Landau-Zener 模型在非线性超冷原子系统中的扩展, 现在称之为非线性 Landau-Zener 隧穿 [44-46]. 描述系统的方程包含了粒子间相互作用, 即方程 (2.6) 中 $c \neq 0$ 的情况. 图 2.1 显示了该模型在耦合 $v = 1.0$ 时的能级结构和隧穿率随扫描速率的变化情况 ($\gamma = \alpha t$, α 为扫描速率), 可以看到非线性相互作用的存在使得系统发生了根本性的变化. 能级结构由于非线性的存在发生了拓扑结构的变化. 没有非线性时, 上下能级结构对称, 见图 2.1(a); 当非线性相互作用超过一个临界值时, 对称性被打破, 下能级出现了一个环状结构, 见图 2.1(b); 这使得系统的绝热性遭到了破坏, 也就是说, 即使绝热极限下, 系统也会有非零的隧穿率, 见图 2.1(c). 有许多文献报道了该现象, 这一结果已经被几个实验所证实 [30,43]. Liu 研究小组在这方面做了大量的工作 [46-51], 也将 Landau-Zener 模型推广到三能级系统 [52], 研究了其隧穿动力学.

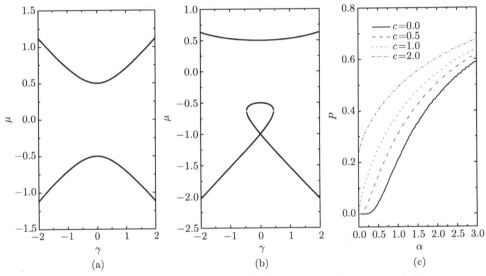

图 2.1 Landau-Zener 模型的能级结构和不同相互作用下的隧穿率

(a) $c = 0$ 时能级结构; (b) $c = 2.0$ 时能级结构; (c) 不同相互作用下隧穿率

Landau-Zener 模型在非线性方面的另一个拓展是: 能级扫描不是线性的, 而是非线性的 (不考虑粒子间相互作用). 一个有趣的现象是在低速扫描范围, 隧穿率会发生振荡, 出现了干涉 [53].

关于 Josephson 振荡, 2005 年, 该现象和自囚禁现象首次在两阱玻色--爱因斯坦凝聚系统中被观察到 [54], 实验观测结果如图 2.2 所示, 左右阱中布居数的时间演化可直接从吸收图像中看出来. 图 2.2(a) 中, 初始的布居数差明显低于自囚禁的临界值, 很明显可以看到原子在两阱之间隧穿, 这种非线性 Josephson 振荡的周期约

2.1 非线性隧穿动力学

为 40ms, 比非相互作用的原子在同样势阱中的振荡周期 (约 500ms) 要小得多. 这一点体现了玻色–爱因斯坦凝聚中原子间相互作用会对系统的动力学行为产生重要影响. 图 2.2(b) 中, 初始的布居数差高于自囚禁的临界值, 可以看到, 双阱中的原子布居数差几乎不随时间变化, 即出现了宏观自囚禁现象. Fu 和 Liu[55] 等从两模量子哈密顿出发, 考虑了粒子间相互作用, 研究了这些现象及其背后的相位问题, 并研究了三势阱中 Josephson 振荡和自囚禁现象 [56].

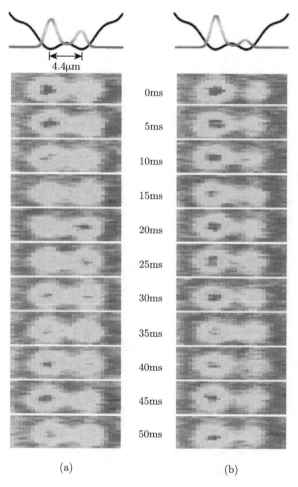

图 2.2 对称双势阱中弱耦合的两阱玻色–爱因斯坦凝聚的隧穿动力学
(a) Josephson 振荡; (b) 宏观量子自囚禁 [54]

关于 Rosen-Zener 隧穿, Ye 等 [57] 将其推广到非线性情况, 研究了非线性相互作用对体系隧穿动力学的影响, 发现了许多新奇的现象, 在玻色–爱因斯坦凝聚和固体物理中有着极好的应用,并利用该过程研究了物质波的干涉,设计了基于玻色-爱因斯坦凝聚的量子干涉仪[58,59].

关于其他模型在非线性方面的推广,主要体现在超冷原子-分子转化问题上(包括前面的 Landau-Zener 模型和 Rosen-Zener 模型). 超冷原子-分子转化系统(关于超冷原子-分子转化将在 2.3 节专门介绍)本身就是非线性的(原子模和分子模相互耦合). Ishkhanyan 等[60-62]在这方面做了大量的工作.

尽管人们对这些模型在非线性方面的推广做了不少出色的工作,然而,在超冷原子系统中,这些模型所对应的一系列现象仍然是未知的. 例如,在 Landau-Zener 隧穿中,如果能级差不是线性扫描,而是非线性扫描,或者能级差和耦合都随时间变化,如 Nikitin 模型和 Demkov-Kunike 模型或其他模型,这在实际的物理系统中是常见的,特别是在具有粒子间相互作用的非线性系统中会发生什么现象,是值得研究的问题. 本书第 3 章将专门对包含粒子间相互作用的系统进行研究.

2.2 高保真度量子操控

2.1 节对非线性隧穿动力学做了简要介绍,在量子系统的隧穿动力学研究中,还有一类问题越来越引起科研工作者的兴趣,即高保真度量子操控. 本节对高保真度量子操控进行简要介绍.

高保真度是量子操控中的一个基本要求,特别是在量子计算中,保真度的要求非常苛刻,其容许误差应该控制在 10^{-4} 以内[63]. 近些年来,围绕高保真度这一核心,在隧穿动力学方面,人们做了许多工作,提出了一系列行之有效的方法[64].

2.2.1 超绝热量子操控

量子绝热定理表明要使系统绝热跟随,系统参数是慢变的,这个过程需要很长的时间,这在实际中是不利的. 因此,保证很好的稳定性和高保真度的同时,需要加速绝热过程. 超绝热技术也是实现高保真度加速绝热量子控制的一种有效方法. 该方法最早由 Berry[65], Lim 和 Berry[66], Demirplak 和 Rice[67,68] 率先提出. 通过在系统中增加一个附加哈密顿量去抵消系统中的非绝热振荡,从而实现高保真度量子操控,后来采用 Berry 的命名,人们称之为超绝热技术. Chen 等提出的绝热捷径技术,利用 Lewis-Riesenfeld (LR) 不变量来实现量子态操控,也能实现绝热加速[69-72]. 尽管这两种方法形式上不同,但是本质上是等价的[72].

近几年,这些技术已经用在许多量子系统中[73-84]. 实验上,超绝热技术已经在加速光晶格的玻色-爱因斯坦凝聚体系[85,86]、钻石单 NV (nitrogen-vacancy) 色心的电子自旋系统[87]、大单光子失谐的冷原子系统[88]、俘获离子的绝热输运的连续变量系统[89] 以及固态 Lamda 系统[90] 中得以实现,证明了该方法的可行性和有效性. 同时, 也有研究者对超绝热技术做了进一步推广. 例如, Baksic 等 [91] 提出的缀饰态 (dressed states) 方法以及 Chen 等 [92] 提出的更普通的附加哈密顿量绝热捷径方法, 都能实现加速绝热过程, 而且在特殊情况下, 都能变为超绝热技术.

2.2.2 超快量子操控

高保真度超快量子驱动是量子操控领域的又一重要内容. 日常生活中, 人们常常关心如何以最快的速度到达某个地方, 在科学研究中, 这种以"最快"为目标的问题就称为"时间最优"问题. 该研究源于三百多年前约翰·伯努利提出的最速降线问题. 量子系统中, 由于量子退相干效应 [93,94], 量子控制过程要尽可能比典型的退相干时间快. 时间最优问题就聚焦于如何在最短时间内将一个量子系统驱动到目标态, 从而实现量子速度极限 (quantum speed limit) 问题, 达到时间最优的量子操控. 早在 1927 年 Heisenberg[95] 就研究这一问题, 即著名的量子力学能量时间不确定关系. 1945 年, Mandelstam 和 Tamm [96] 获得了一个量子速度极限不等式, 之后许多学者分别在不同的物理系统 [97,98], 包括开放系统、非马尔科夫系统、多体系统以及相对论量子系统中研究了该问题 [99-102], 建立了量子速度极限优化控制等理论 [103].

近十几年来, 时间最优量子控制的量子速度极限问题也越来越引起人们的广泛关注. 近些年来, 量子最速降线方程的提出为一大类时间最优量子控制问题的研究提供了理论框架. 2015 年, 美国 MIT 的 Seth Lloyd 研究组基于多比特量子系统发展了求解量子最速降线方程的数值方法. 中国科学技术大学在此方面研究工作出色, 不断有新成果出现, 2014 年实现了精度高达 0.996 的单比特量子操作 [104], 随后实现了达到容错量子计算要求 (单比特操作精度达到 0.999952, 两比特操作精度达到 0.992) 的普适量子逻辑门 [105]. 2016 年, 基于具体的量子物理体系, 进一步发展了实现普适量子控制的时间最优控制方法, 如图 2.3 所示, 并在固态自旋系统金刚石 NV 色心体系中得以实现 [106]. 该工作表明, 其量子操作精度高、耗时少, 从而证实了以时间最优的方式实现精确量子控制的可行性, 这也意味着在单位时间内可以实现更多的高精度量子逻辑门, 因而在未来量子计算等领域具有非常重要的应用前景. 在超冷原子系统中, Bason 等 [85] 实现了光晶格玻色-爱因斯坦凝聚体中的超快量子驱动问题, 发现复合脉冲技术能实现时间最优量子控制, 实验示意图如图 2.4 所示. 图 2.4(a) 为 Landau-Zener(LZ) 两能级系统, 其中 E_{rec} 是系统的自然能量尺度, 非绝热态 (裸态)$|0\rangle$ 和 $|1\rangle$ 被耦合到绝热态 $|\psi_g\rangle$ 和 $|\psi_e\rangle$. 非绝热能级在 $\tau = 0.5, E = 0$ 处交叉, 箭头表示两种极端技术, 其中水平箭头表示超快量子驱动技术 (沿绝热能级箭头表示超绝热技术). 图 2.4(b) 表示在玻色-爱因斯坦凝聚中的实验实现, 凝聚体囚禁在光偶极阱中并装载在光晶格里. 图 2.4(c) 为凝聚体在光晶格中的能带结构, 盒子表示用来实现在最低的两个能带中等效两能级系统的准动量范围.

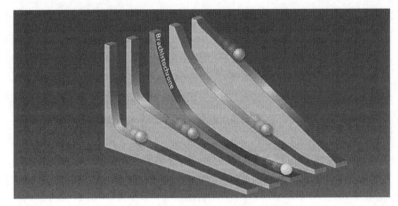

图 2.3 时间最优量子控制 [107]

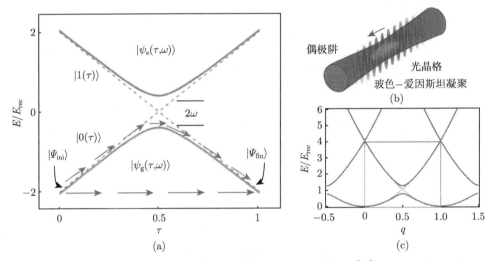

图 2.4 两能级量子系统示意图和实验实现 [85]

(a) LZ 两能级系统; (b) 在玻色–爱因斯坦凝聚中的实验实现; (c) 凝聚体在光晶格中的能带结构

2.2.3 复合绝热通道技术

Vitanov 研究组提出的复合绝热通道技术 [108] 是提高保真度的量子操控方法之一, 该方法基于核磁共振和通常的绝热通道技术, 用一系列具有特定相位的复合脉冲去代替单个脉冲, 通过控制这些脉冲的相位, 从而实现高保真度量子操控, 近几年已经在实验上验证了这一方法的有效性 [109,110]. 该技术拥有共振脉冲和绝热通道技术各自的优点, 既具有高保真度, 同时对实验参数有很好的鲁棒性, 成为量子操控中的一个重要手段.

Daems 等 [111] 提出的单束整形脉冲 (single-shot shaped pulse) 技术, 也需要通

过控制相位进行量子操控, 其相位可根据不同鲁棒性展开成傅里叶级数的形式, 通过确定展开系数来实现操控, 该方法只需要一束脉冲即可, 只是不同鲁棒性要求其脉冲形式略有不同, 其保真度也很高 [110].

总之, 关于高保真度量子操控的研究, 在高精密光谱、分子动力学相干控制、量子态制备、化学反应、原子钟的设计, 特别在量子信息等领域有着极为广泛的应用前景 [110]. 本书第 4 章专门讨论这方面的问题.

2.3　超冷原子–分子转化

超冷分子的研究是超冷原子系统量子操控中的又一重要内容, 是当前物理和化学领域人们最感兴趣的热点研究课题之一. 它涉及少体物理学、化学、精密光谱学、多体物理学、粒子物理学, 甚至天体物理学等领域 [112-118]. 超冷分子的产生为研究分子物质波、分子玻色–爱因斯坦凝聚、强相互作用超流、高精分子谱、相干分子光学以及量子计算机等提供了平台 [119-121]. 超冷原子、分子有着独特的优点, 可观测相干的物质波波长, 具有精确的能级结构, 能在微观尺度上操纵原子、分子, 实现量子态操控, 可用于制造原子干涉仪、原子激光器, 研制冷原子钟、超冷温度计 [122,123], 可在低温下控制化学反应 [124,125], 可用于研究量子多体系统等, 在许多领域有着潜在的应用价值. 开展超冷分子的研究对推动量子计算技术的发展、物理学基本常数的精确测量、分子光谱及结构的精确测量、量子电子器件的制备等都具有十分重要的意义. 本节对超冷分子的有关内容做一个简要的小结.

2.3.1　超冷分子性质

相对于原子而言, 分子结构更加复杂, 自由度更多, 具有平动、振动和转动自由度, 也更加有趣. 图 2.5 表示分子的众多自由度及其相应的能量尺度. 当温度降低时, 各种自由度的活性将减弱 (甚至被冻结). 在超低温下, 分子的振动自由度不活跃 (在一定程度上被冻结), 但电子自旋和核自旋不能被冻结, 可以近似把分子视为刚性转子. 当分子气体被冷却到极低温度时, 它的内部自由度的激发能比它的动能大几乎 10 个数量级 [126,127]. 因此, 精密地控制分子的内部量子态以及内部态和质心运动之间的能量交换, 对于实验的成功极其重要. 这也使得冷却分子的研究极其困难. 与此同时, 也正是由于这些丰富的内部结构, 一旦人们成功地制备出超冷分子气体, 会为精密测量、量子科学和超冷化学等研究提供更多的机遇和挑战 [127]. 分子具有电偶极矩, 因而具有原子所不具备的性质, 从而为一些新的量子操控奠定基础, 如上面提到的量子计算 [121]. 也可以探索物质的一些新行为, 如通过电偶极矩相互作用形成玻色–爱因斯坦凝聚 [128].

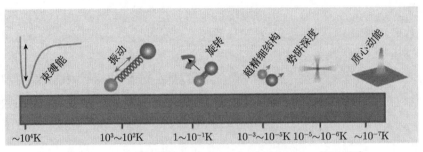

图 2.5 分子的众多自由度及其相应的能量尺度 [126, 127]

从运动速度来看, 随着温度的不断降低, 运动速度也会减小, 这就意味着平动自由度变得不活跃. 室温时, 气体原子和分子一样具有很高的平均速度, 比子弹还快. 然而, 随着环境温度的降低, 其速度也就越来越慢了. 以氮分子为例, 在室温 (300K) 下, 分子的运动速度在 10^2m/s 量级, 和子弹运行速度可比, 而在 10^{-6}K 的超低温下, 运动速度就变成一个 "沉重的蚂蚁" 了, 大约 0.03m/s, 即每秒钟运行几厘米. 如此低温下, 分子和原子一样, 除了速度低外, 还有一个显著的特点: 波动性凸显. 具有波动性的波称之为物质波, 也就是前面提到的 de Broglie 波. 其波长如前所述, 定义为 $\lambda_{\rm dB} = h/\sqrt{2\pi m k_B T}$. 可以看到, de Broglie 波长取决于粒子的质量和温度. 粒子的质量越小, 温度越低, 其波长就越长, 波动性就越明显. 同样以氮分子 (质量 28g·mol^{-1}) 为例, 室温时 (300 K), 波长为 2×10^{-11}m (即 0.2Å), 这个值小于分子的大小 (氮气分子的束缚长度大约是 1.1Å). 当温度冷却到 1K 时, 波长增加到 3.3Å, 和分子大小幅度 (不随温度而变) 相同. 当温度达到 1mK 时, 波长为 3464Å, 即 0.35mm, 和室温时相比已经扩大了 3464/0.2 = 17320 倍. 因此, 通过冷却 (变慢), 室温时的分子抛射物 (molecule-projectile) 已经变成微开温度时的分子波 (molecule-wave), 其波长已经超过分子维数. 图 2.6 显示了 Cs$_2$ 分子 de Broglie 波长随温度的变化情况, de Broglie 波长的单位为玻尔半径, 图中也标示了一些典型现象发生时对应的温度. 超低温下, 分子的动能很小 (小于分子超精细相互作用的能量), de Broglie 波长远远大于分子的典型尺度, 研究此温度领域的化学反应会有一些特殊的应用价值 [129]. 而图 2.7 则显示了不同分子 de Broglie 波长随温度和质量的变化情况, 也显示了不同温度范围获得分子的一些典型技术手段. 通过观测超冷分子 de Broglie 波长可以研究物质波的干涉、衍射等传播特性.

2.3.2 超冷分子的几种产生方法

尽管超冷分子的应用比超冷原子更加广泛, 然而, 分子要冷却下来比原子要困难得多, 如何获得超冷分子呢? 本节简单介绍超冷分子形成的几种方式.

总体而言, 要形成超冷分子可分为直接方法和间接方法两种 [118, 130].

2.3 超冷原子–分子转化

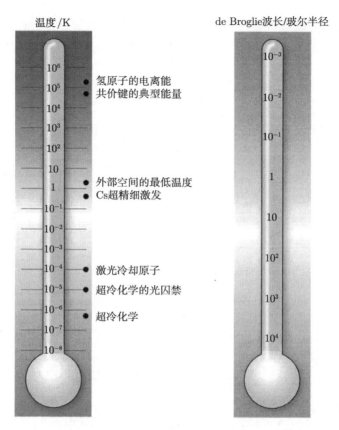

图 2.6 Cs_2 分子 de Broglie 波长随温度的变化情况 [129]

直接方法就是将原先存在的分子直接冷却 (变慢). 要么通过随时间变化的电场、磁场或者辐射场去减速超音速的分子束, 要么从分子束中挑出慢的分子, 主要包括: 缓冲气体冷却[131]、Stark 减速[132]、脉冲光场减速[133,134]、交叉分子束的碰撞减速[135,136]、中心旋转喷嘴的超声速扩展技术[137] 以及从麦克斯韦–玻尔兹曼分布的分子束中挑选低速分子[138] 等. 所有这些直接方法开始都是以相对热 (200~1000K) 的分子为对象, 通常先形成超声速的分子束. 这些方法的优点是: 可以应用于许多种类的分子 (如 Stark 减速能用到所有的极性分子, 缓冲气体冷却和磁俘获相结合可用于所有的顺磁性分子). 以上方法中, 相空间密度保持为常数. 更令人兴奋的是, 2010 年 10 月, 美国耶鲁大学物理系 Shuman 等[139] 在实验上首次实现了双原子极性分子 SrF 的激光冷却 (毫开量级). 再如, 同年, 美国 JILA 实验小组利用超冷化学反应形成超冷分子[124,125], 如图 2.8 所示, 两个 KRb 分子在低于 1μK 温度下发生化学反应形成 K_2+Rb_2. 这些研究无疑为超冷分子的产生提供了另外一些途径. 表 2.1 显示了超冷分子研究中不同温度范围冷原子气体相空间密度、粒子数密度和对应的主要科学目标, 表中分子偶极矩假定为 1 Debye, 分子质量假定为 100 u. 而图 2.9 显示不同温度、密度下形成冷和超冷分子的技术方法及其对应的科学应用.

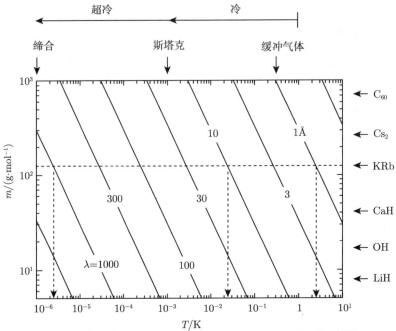

图 2.7　不同分子 de Broglie 波长随温度和质量的变化 [118]

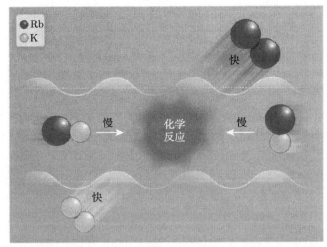

图 2.8　量子操控的冷化学反应: 两个 KRb 分子在低于 1μK 温度下发生化学反应形成 K_2+Rb_2 [125]

2.3 超冷原子-分子转化

表 2.1 相空间密度、粒子数密度和对应的主要科学目标[116]

相空间密度/$\hbar^3$	粒子数密度/cm^{-3}	温度	科学目标
$10^{-17} \sim 10^{-14}$	$10^6 \sim 10^9$	<1K	自然界中基础力的检测
10^{-14}	$>10^9$	<1K	电偶极相互作用
$10^{-13} \sim 10^{-10}$	$>10^{10}$	<1K	冷控化学
10^{-5}	$>10^9$	<1μK	超冷化学
1	$>10^{13}$	100nK	分子的量子兼并
1	$>10^{13}$	100nK	分子光晶格
10	$>10^{14}$	<100nK	新奇量子相变
100	$>10^{14}$	<30nK	偶极晶体

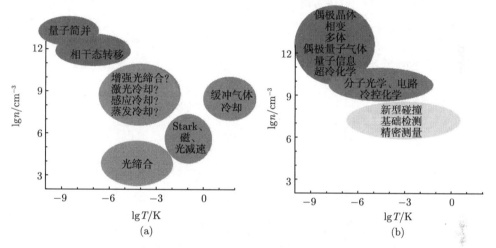

图 2.9 不同温度、密度下形成冷和超冷分子的技术方法及其对应的科学应用

(a) 不同空间密度和温度时形成冷和超冷分子的技术方法; (b) 一定密度和温度下冷和超冷分子在实际中的应用 (超冷分子的重要科学意义) [116]

间接方法一般是指将原先的超冷原子采用各种方法转化为超冷分子. 其中, Feshbach 共振、光缔合和受激拉曼绝热技术是直接实现从超冷原子到超冷分子转变的重要工具, 利用这些技术可以将原子玻色-爱因斯坦凝聚体, 甚至简并费米子-费米子混合物或玻色子-费米子混合物转化为双原子分子或者更复杂的分子凝聚体. 以下分别对这些技术进行简要的介绍.

1. Feshbach 共振

Feshbach 共振是美国物理学家 Feshbach (1917~2000 年) 在研究热中子重核散射中于 20 世纪 50~60 年代发现的一种量子散射现象[140,141]. 20 世纪 90 年代, Tiesinga 等[142] 预言在碱金属原子气体中存在该现象, 提出该系统中可通过改变电磁场来调节原子碰撞的散射长度. 1998 年 MIT 的 Ketterle 研究组首先在钠原子系统的实验中观测到 Feshbach 共振[143]. 2003 年, Feshbach 共振被用于产生超冷分子, 获得了许多种类的超冷分子. 利用磁场调整原子之间的相互作用, 使一对散射态的原子可以形成一个弱的束缚态, 即碰撞中的两个原子可以直接整合变成一个分子.

Feshbach 共振的基本原理是: 两个具有特定能量的粒子在散射过程中能耦合到一个量子束缚态. 在超冷气体物理学中, 这种碰撞过程发生在非常低能量的散射过程中, 当两个自由碰撞的原子 (称为原子散射态, 三重态, 开道) 耦合到一个分子态 (称为束缚态, 单重态, 闭道) 时会发生. 通常原子散射态和分子束缚态拥有不同的磁矩, 原子散射态和分子束缚态之间存在能量差, 通过外加磁场可调节这个能量差, 在某个特定的磁场, 两态调至相同的能量, 散射长度 a 发散 (发生共振), 原子间相互作用加强 (散射长度和原子间相互作用成正比), 从而产生 Feshbach 共振. 其原理图如图 2.10 所示.

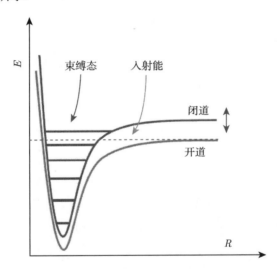

图 2.10 Feshbach 共振发生过程示意图[114]

s 波散射长度是描述超冷碰撞过程的一个关键参数, 它完全特征化散射性质. Feshbach 共振时散射长度满足下列关系:

$$a = a_0 \left(1 - \frac{\Delta}{B - B_0}\right), \tag{2.11}$$

其中, B_0 为共振发生时磁场的位置; a_0 为背景散射长度 (远离共振时的散射长度), 远离共振时, $a \to a_0$; Δ 称为共振宽度. 在不同的系统中, 该宽度值从几毫高斯到几百高斯. 根据共振宽度的大小, 通常将 Feshbach 共振分为窄共振 (narrow resonance)

2.3 超冷原子–分子转化

和宽共振 (broad resonance), 一般来说, 当 $\Delta < 1\mathrm{G}$ 时, 称为窄共振; 当 $\Delta > 1\mathrm{G}$ 时, 称为宽共振[117]. 散射长度和原子–原子形成两体的结合能 E_b 紧密相关:

$$E_\mathrm{b} = -\frac{\hbar^2}{ma^2}, \qquad (2.12)$$

其中, m 是原子质量. 可以看到, 当散射长度增加到无穷大时, $E_\mathrm{b} \to 0$, 在大的负值时, $E_\mathrm{b} > 0$.

图 2.11 所示为 Feshbach 共振附近原子散射态和分子束缚态耦合, 以及散射长度和散射截面的变化情况. 当磁场扫描穿过 Feshbach 共振 (箭头) 时, 原子和分子耦合, 在原子的散射过程中, s 波散射长度或散射截面在 Feshbach 共振附近会产生相当剧烈的变化, 展现出很强的共振行为. 三体碰撞过程在 Feshbach 共振附近也显著增强. 当分子束缚态能量略低于散射态能量时, s 波的散射长度为正值, 而当分子态能量略高于散射态能量时, 散射长度为负值. 在共振状态时散射长度会发散, 同时弹性散射截面也会共振式地增强. 共振的宽度则取决于耦合强度和原子与分子的磁矩差. 原子玻色凝聚实验中, 原子间有效相互作用 (也称为平均场相互作用) 由其散射长度所决定. 当散射长度为负值时, 玻色原子间相互作用是相互吸引的, 从而导致玻色–爱因斯坦凝聚塌陷; 当散射强度为正值时, 原子相互作用力表现为排斥, 玻色–爱因斯坦凝聚是稳定的. 通过 Feshbach 共振, 可以任意地调整量子气体之间的相互作用.

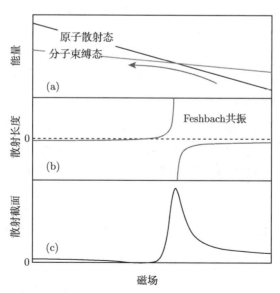

图 2.11 Feshbach 共振
(a) 共振附近原子散射态与分子束缚态的耦合; (b) 散射长度随磁场的变化;
(c) 散射截面随磁场的变化[144]

同样, 可以降低磁场穿过共振将散射中的原子通过绝热过程转变成分子, 这种动态通过共振产生超冷分子的方法称为绝热的 Feshbach 扫描 (Feshbach ramp) [144-146], 所产生的分子一般称为 Feshbach 分子, 这些分子的大小通常在几十到几百玻尔半径 (1 玻尔半径等于 0.05nm).

利用 Feshbach 共振技术, 从超冷原子产生超冷分子的方法有三种 [117], 分别如图 2.12 所示. 图中实线表示弱束缚的分子态, 在共振点 $B = B_0$ 时能够离解成连续的原子态 (点线).

(1) Feshbach 扫描. 扫描磁场通过 Feshbach 共振, 由于能级免交叉效应 (avoided level crossing), 两个相互作用的原子可以在绝热状态中结合成一个分子.

(2) 振荡磁场 (oscillatory magnetic fields) 方法. 其产生超冷 Feshbach 分子的方法是基于调制的振荡磁场, 振荡磁场诱导两个碰撞的原子形成束缚态分子. 由于缔合过程发生在远离共振位置处, 因此会降低热和原子损失, 提高分子转化效率.

(3) 三体重组 (three-body recombination) 过程. 根据能量和动量守恒原理, 二体碰撞无法产生稳定的分子, 在 Feshbach 共振附近, 三体碰撞过程可以使其中两个原子结合成一个稳定分子 [117,146].

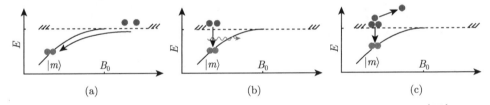

图 2.12 磁场 Feshbach 共振在实验上产生超冷分子的三种技术示例图 [117]

(a) 磁场扫描通过共振点时两个相互作用的原子绝热转化成一个分子; (b) 振荡磁场驱动由散射态向分子态转变; (c) 三体重组形成分子

采用 Feshbach 共振方法, 早在 2002 年, Donley 等 [147] 在 ^{85}Rb 玻色–爱因斯坦凝聚中观察到原子对与分子的相干振荡, 此后许多研究小组分别实现了 ^{133}Cs$_2$ 分子 [148-150]、^{87}Rb$_2$ 分子 [149]、^{23}Na$_2$ 分子 [150] 等. 对于费米原子而言, 已经实现的超冷分子有: ^{40}K$_2$ 分子 [151]、^{6}Li$_2$ 分子 [152-154] 等. 近年来, 也实现了异核分子, 如 ^{40}K^{87}Rb [155-157]、^{6}Li^{23}Na [158]、^{6}Li^{40}K [159]、^{85}Rb^{87}Rb [160]、^{41}K^{87}Rb [161] 等.

Liu 小组在 Feshbach 共振方面的工作主要集中在探讨超冷原子–分子转化中的多体效应以及粒子间相互作用对转化过程的影响, 定性和定量解释一些著名的超冷分子转化实验 [36,162,163], 并研究了转化过程中出现的一些奇特现象 [164].

由 Feshbach 共振产生的分子, 处于高激发振动能态且具有相当小的束缚能. 内在的能量可在原子–分子或分子–分子的碰撞过程中经由振动内能耗散的方式释放出来, 这些过程会导致在光阱中的分子数目快速减少. 因此, 对于由原子玻色–爱因

2.3 超冷原子-分子转化

斯坦凝聚所形成的分子, 上述非弹性碰撞过程会导致超冷分子的寿命极短, 被限制在数个厘秒 (10^{-2}s) 范围内. 然而, 对于费米系统, 由 Feshbach 共振产生的分子具有非常优越的碰撞稳定性来抵制非弹性散射过程. 对于一般的费米原子气体, 因为泡利不相容原理无法任意靠近, 然而对于组成分子的费米原子气体, 在分子经过碰撞而产生非弹性散射的过程中, 这些组成分子的费米原子必须更加靠近, 因而分子在碰撞过程中产生有效的斥力, 并导致它们具有好的稳定性. 因此, 对于弱束缚的"费米原子对"而言, 在低温下其寿命可以比玻色原子组成的分子的寿命长许多个数量级, 能达到数秒. 所以, 对于费米系统, Feshbach 共振是用于产生超冷分子的一种非常有效的技术 [146,165]. 这些分子稳定性的研究使分子玻色-爱因斯坦凝聚的形成有了可能, 同时, 如果由费米原子耦合而成的玻色分子寿命足够长, 就可以通过降低体系的温度来实现分子玻色-爱因斯坦凝聚 [145]. 2003 年 Jin 小组从囚禁在光偶极势阱中深度简并的 ^{40}K 费米原子气出发, 采用 Feshbach 共振扫描技术获得了超冷玻色分子, 并进一步通过降低势阱的深度对分子进行蒸发冷却, 从而实现了 ^{40}K$_2$ 的分子玻色-爱因斯坦凝聚 [151], 见图 2.13. 同时 Innsbruck 的 Grimm 团队和 MIT 的 Ketterle 团队也成功实现了 ^{6}Li 费米原子的分子玻色-爱因斯坦凝聚 [146,165].

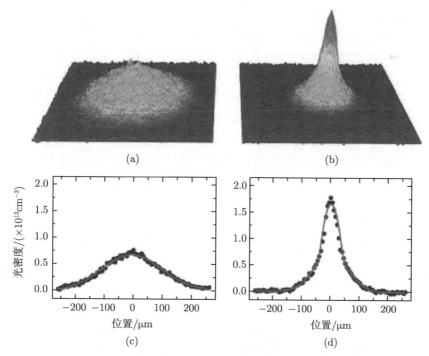

图 2.13 钾分子云在凝聚前 (a)、(c) 和凝聚后 (b)、(d) 的密度分布 [151]

2. 光缔合

另一种有效产生超冷分子的方法是光缔合 (photoassociation, PA) 技术. 该技术早在 1987 年由 Thorsheim 等[166] 预言. 目前已经有许多综述文章[113,167]. 光缔合是指两个碰撞的原子吸收一个光子形成一个激发态分子的过程. 早期有关这方面的工作主要聚焦于单光子光缔合, 形成激发态分子. 近几年, 光缔合技术已经利用双光子过程用于形成电子基态超冷分子. 利用该技术, Fioretti 等[168] 首次在实验上实现了基态超冷 Cs_2 分子, 后来 Nikolov 等[169,170] 也实现了基态超冷 K_2 分子, 同时也实现了超冷异核分子, 如 $^{85}Rb^{133}Cs$[171,172] 和 $^{39}K^{85}Rb$ 分子等, 温度在 $100\mu K$[173].

光缔合技术实现冷原子到冷分子转化如图 2.14 所示. 图 2.14(a) 表示两个自由原子在碰撞过程中吸收了激光光子而形成一个分子态. 由于光缔合形成的分子能保持自由原子的温度, 因此从冷原子形成冷分子是可能的. 图 2.14(b) 显示了原子基态和激发态的势能曲线随原子核间距离的变化情况, 也显示了分子基态和激发态的振动能级. 一对基态原子碰撞不能掉进分子基态势阱中, 然而, 可以产生一个束缚激发态: 它足以激发碰撞原子到和上面势能曲线相关联的振动态. 这个态有比电子激发态阱深低的能级 (注意这里冷原子的平动能量和振动能量相比可以忽略). 形成束缚激发态并不意味着就此结束, 而是一个中间步骤, 这样一个态仅仅是一个亚稳态, 能自发辐射 (典型的时间是微秒), 要么返回到最初的基态原子, 要么到分子态, 形成的分子处在一个高振动激发态 (一个极化分子振动态相对寿命是 $10 \sim 100ms$, 非极化分子在不存在外部扰动时寿命是无限长的). 这样形成的振荡激发态分子有一个问题: 如果这样的分子碰撞, 它们的振动能量就能够转化成平动能量, 分子会变热 (K 或者更高), 这样就达不到光缔合形成冷分子的目的. 过去一些年里, 通过另外一个附加激光场 (双色光缔合) 使激发态受激跃迁成为可能, 这样实现了从激发态到基态的布居数反转, 见图 2.14(c), 从而形成超冷基态分子. 该技术和 "相干布居数转移" 原理相同.

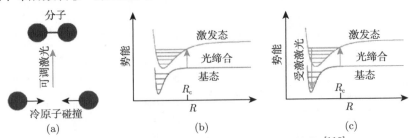

图 2.14 光缔合技术实现冷原子到冷分子转化[118]
(a) 光辅助的碰撞过程; (b) 基态和激发态势能曲线, R_c 表示原子核间距离;
(c) 激光受激辐射基态和激发态势能曲线

2.3 超冷原子-分子转化

由以上分析可知, 采用 Feshbach 共振和光缔合技术产生的分子往往处于不稳定的状态. 对于费米系统而言, 库仑阻塞作用抑制了分子的衰减, 这两种技术是可行的. 然而对于玻色系统, 产生的分子很快会发生离解, 有大量的非弹性损失, 从而大大降低了超冷分子的转化效率[165]. 为了得到更高的从超冷原子到稳定紧束缚分子的转化效率, 光缔合受激拉曼绝热技术作为一种更有效的方法被提出来, 下面简要介绍这方面的工作.

3. 受激拉曼绝热过程

受激拉曼绝热通道 (STIRAP) 技术是一种更加有效的产生超冷分子的方法. 该方法的成功主要依赖于存在相干布居数俘获 (coherent population trapping, CPT) 态, 即暗态. 在暗态上, 激发态上的粒子数布居没有分布, 从而使得暗态上的绝热演化有效抑制了激发态上粒子数的自发辐射, 实现高的分子转化效率[165]. 该技术于 2000 年由 Mackie 等[22] 首先预言可以用来实现超冷原子-分子的转化.

以三能级系统为例, 受激拉曼绝热通道技术如图 2.15 所示. 从图 2.15 可以看到, 三能级系统包含三个非简并态: 初态 $|1\rangle$、中间态 $|2\rangle$ 和终态 $|3\rangle$, 分别对应于三个不同的能级. 初态 $|1\rangle$ 和终态 $|3\rangle$ 分别由 Stokes 激光 S 和 Pump 激光 P 通过中间态 $|2\rangle$ 耦合. 中间态能通过自发辐射等衰变到其他的能级. 从跃迁频率到中间态 Pump 和 Stokes 激光频率失谐分别是 Δ_P 和 Δ_S. 允许的偶极跃迁有 $1 \leftrightarrow 2$ 和 $2 \leftrightarrow 3$. 两束相干激光用于诱导从初态到终态的转移, Pump 激光用于耦合初态和中间态, Stokes 激光用于耦合中间态和终态. STIRAP 技术的目标是通过操控外场来控制三个态的布居数分布, 使得中间态 $|2\rangle$ 上最终没有布居数分布, 这样就避免了中间态 $|2\rangle$ 的自发辐射等损失, 实现了从初态 $|1\rangle$ 到终态 $|3\rangle$ 的理想传输, 这种不包含高激发态的相干态就成为相干布居数俘获态, 即暗态. 对于原子-分子转化系统而言, 三

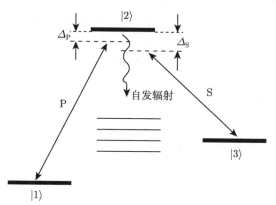

图 2.15 三能级系统受激拉曼绝热通道技术示意图[174]

个态分别是初始原子基态、中间分子激发态和最终分子基态,利用该技术使得激发态分子布居数为零,从而获得基态原子和终态分子基态的相干叠加暗态,实现稳定且高的原子–分子转化效率.

大量的工作围绕该技术探讨超冷原子–分子的转化[165]. 实验方面, Winkler 等[175]于 2005 年观察到原子–分子暗态,并实现了 Rb 原子玻色–爱因斯坦凝聚到基态超冷 Rb_2 分子的转化. 由于实验中各个能级上粒子的自发辐射,其转化效率较低. 2007 年,该小组利用该技术把光晶格中的超冷 Feshbach 共振弱束缚分子转移到更深的束缚振动量子能级上,转化效率高达 87%,分子寿命为 1s [165,176]. 2008 年, NIST 和 JILA 实验小组利用该方法将 Feshbach 共振产生的弱束缚 $^{40}K^{87}Rb$ 极性分子转移到深束缚振动基态能级上,得到分子温度达 350 nK [165,177]. 在理论方面,和传统的 STIRAP 方法不同,原子–分子转化系统存在非线性,这些非线性来源于粒子间相互作用的平均场处理以及原子转化成分子的过程. 前者会引起系统动力学的不稳定性,使得暗态不能绝热跟随;后者使得系统不再具有 $U(1)$ 不变性,传统的保真度定义不再适用[165]. 人们已经投入了许多精力研究该技术在原子–分子转化系统中的应用[178-184]. 如 Drummond 等[180]研究了自发辐射和粒子间相互作用对该技术实现原子–分子转化效率的影响,提出了优化的 STIRAP 技术. Mackie[181] 提出了 Feshbach 共振辅助的 STIRAP 技术,产生稳定的超冷分子. Pu 研究小组针对原子–分子转化中的非线性,提出了广义的 STIRAP 技术,利用双光子共振条件来抵消非线性带来的影响,实现了稳定高转化率的双原子分子[182]. 之后,许多工作陆续展开,如利用 Feshbach 共振辅助的 STIRAP 技术去产生更为复杂的超冷同核和异核双原子分子、三原子分子和四原子分子[165,185,186].

Meng 等小组也研究了具有非线性粒子间相互作用的原子–分子转化系统的绝热性和保真性,定义了系统的绝热保真度[146,165,187,188]. 这些工作的开展对实验的顺利进行具有指导意义,也在非线性光学、量子计算等应用学科中具有重要意义.

值得指出的是,不仅可通过以上三种典型方法单独产生超冷分子,近年来人们也发展了许多间接形成超冷分子的方法. 例如,将以上方法适当组合来实现超冷原子–分子转化,如一种组合光缔合和 Feshbach 共振的新方法,也称为优化 Feshbach 的光缔合技术[189-191],被用于产生超冷分子以及上面提到的 Feshbach 共振辅助的 STIRAP 技术等.

如何形成高转化率、稳定的超冷分子是超冷原子–分子转换研究中的永恒课题. 就 Feshbach 共振技术而言,线性扫描磁场通过 Feshbach 共振点会诱导热效应、产生大量的粒子损失,而且形成的分子在高激发态极不稳定. 为此,人们在 Feshbach 共振过程中采用正弦振荡型磁场,或者将 Feshbach 共振和光缔合技术相结合等技术手段来获得超冷分子,提高了分子转化率并降低了分子形成过程中粒子损失. Li 研究组通过在 Feshbach 共振区设计矩形磁场脉冲,提高了分子转化率[192]. 能否在方法上有所创新, 发展高效稳定形成超冷分子的有效技术, 这是值得期待又有实际意义的研究课题. 本书第 5 章将探讨这方面的问题.

然而, 以上所介绍的方法主要用于形成双原子分子, 偶尔形成三原子分子和四原子分子. 多原子分子, 即多聚物分子, 在研究复杂化学反应 (包括超冷化学和超化学) [193,194]、分子退相干 [195]、精密测量 [196] 以及分子光学 [197] 等方面具有重要意义. 那么, 如何形成多聚物分子? 2.3.3 节将从近几年来的一个热点研究领域 —— Efimov 共振入手, 简要介绍超冷多聚物分子的形成方法.

2.3.3　Efimov 共振及其超冷多聚物分子的形成

Efimov 共振是近几年来量子物理领域中极其热门的研究课题之一, 它为研究量子系统中少体物理开启了一扇巧妙的窗户. 该现象于 1970 年由俄罗斯物理学家 Efimov 预言 [198], 并于 2005 年由奥地利因斯布鲁克大学以 Grimm 教授为首的研究组在 10nK 的超低温度下, 首次在铯原子气体中被观察到 [199].

20 世纪 70 年代, 年轻的 Efimov 在 Leningrad 核物理研究所期间, 由于受到 1935 年 Thomas[200] 关于氚核 (triton) 三体问题的启发, 对具有共振两体相互作用情况下的三个全同玻色子体系产生了浓厚的兴趣. 他发现了一个奇特的量子现象: 三个全同的玻色子在具有大的两体散射长度 (即具有共振两体相互作用) 情况下, 即使不存在两体束缚态时, 也会形成多种构型的弱三体束缚态 (无穷多松散的束缚态), 其大小远远超过两体相互作用势的特征尺度范围; 并且不论它们的组分和相互作用的宏观特性如何, 这样的态还具有一些普适的行为 (universal behaviour), 如能量谱的几何尺度率. 后来人们称这种现象为 Efimov 效应或者 Efimov 共振, 将产生的三体态称为 Efimov 态.

Efimov 共振的基本物理图像如图 2.16 所示. 图中画出了三体系统能量谱随散射长度倒数 $1/a$ 的变化情况. 对应的阈值将平面分成三部分: $E > 0$ 区域对应的是三原子连续态, $a > 0$ 并且 $-\hbar^2/(ma^2) = E_b < E < 0$ 区域对应的是原子-两体连续态以及三体态区域. Efimov 态出现在 $a < 0$ 的三原子阈值和 $a > 0$ 时原子-两体态阈值以下的范围. 箭头表示 Efimov 三体态和三原子阈值以及原子-两体态阈值的交点, 意味着三体 Efimov 态耦合到这个阈值时会导致共振现象发生 [201]. 也就是说, $E > 0$ 区域系统由具有一定动能的三个自由原子组成, 零能量阈值 (水平实线表示, $E = 0$) 以及 $E < 0$ 区域涉及束缚态. $a > 0$ 时, 存在一个两体弱束缚态, 在共振范围 (共振极限时 $1/a \to 0$), 其束缚能为 $E_b = -\hbar^2/(ma^2)$, 对应于图中的抛物线, 两体态的存在导致了原子-两体态阈值, 阈值之上两体态与一个自由原子共存. 最为有趣的范围是 Efimov 态存在的区域, 只有当 $a < 0$ 时在三原子的阈值之下, 并且在 $a > 0$ 且原子-两体阈值之下时, 三体束缚态才存在. 当 $1/a$ 在负值一边趋于共振区域时, 在不存在弱束缚两体态时第一个 Efimov 三体态出现了, 当经过共振

时该态到达正 a 一边, 最终和原子–两体态阈值相交融.

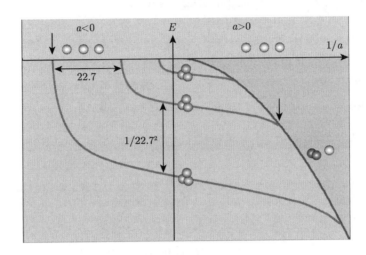

图 2.16　Efimov 共振的基本物理图像: 两个可以"见异思迁", 三个却能"团结一心"[201]

Efimov 发现从 $a<0$ 时三原子阈值到 $a>0$ 时原子–两体态阈值范围有一系列这样的三体态存在. 同时, 也发现了一个令人惊讶的几何尺度率, 包含因子 22.7 (这个值和 e^π 有一点小的差别, 图 2.17 表示的是俄罗斯套娃 (Russian nesting dolls), 其几何尺度率是 1.3). 如果将散射长度增加 22.7 倍, 另一个三体束缚态就会出现, 其大小是原来的 22.7 倍, 而能量减小到原来的 $1/22.7^2$. 即散射长度以尺度率 22.7 增加, 而束缚能以 $1/22.7^2 \approx 1/515$ 减小时, 另一个 Efimov 三体态会出现. 依次类推, 会出现一系列 Efimov 三体态, 第 n 个 Efimov 态对应的散射长度是第一个的 22.7^n 倍, 大小是原来的 22.7^n 倍, 而能量减小到原来的 $1/22.7^n$.

图 2.17　俄罗斯套娃

此现象进一步引起人们的极大兴趣并且违反人们直觉的性质还有: ① $a < 0$ 时, 三体束缚态 (也称作 Borromean 态, 正如著名的 Borromean 环一样, 要重移其中的任何一个整环就会遭到破坏, 即只要打开其中一个环, 另外两个将不复存在. 如图 2.18 所示, $a < 0$ 的范围也称为 Borromean 范围) 在没有任何两体束缚态存在时也存在. ② $a > 0$ 时, 随着两体束缚强度增强, 三体态会消失. "Borromean" 一词在历史上来自意大利一个叫作 "Borromeo" 的贵族家庭, 他们用这个环作为家族的族徽 (coat of arms).

当时, 科罗拉多大学的 Greene 教授对这个现象的评价是 "不可思议". 由于原来的两个粒子似乎突然接通了开关, 把与它们相距有一定长度的第三个粒子吸引过来, 组成一个三体态. 在随后的十年间, 想用实验来观察 Efimov 共振的工作几乎无进展, 研究者们仅在理论上用不同的方式证明了 Efimov 的预言[202].

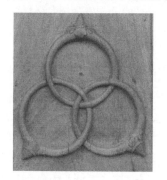 

图 2.18　Borromean 环

1977 年, 氦三体被考虑为一个有趣而基本的观察到 Efimov 态的候选者[203], 大约有 50 多个理论工作致力于该问题的讨论[201], 然而关于分子束的实验不能证实理论预言, 他们要么怀疑氦中不存在 Efimov 态[204], 要么考虑其他的分子作为实现 Efimov 态的候选者[205].

1993 年, Feshbach 共振现象被预言并于 1998 年首次在实验中实现, 由于利用 Feshbach 共振可任意调节粒子间相互作用, 引起了人们在超冷系统中探索 Efimov 态的极大兴趣. 1999 年, Greene 教授与合作者 Esry 和 Burke 等共同提出了一个大胆的设想: 他们认为处于超冷状态下的原子气体中会出现 Efimov 效应. Stanford 的朱棣文研究小组推测在他们的 Cs 原子实验中可能会出现 Efimov 态[206], 实际上, 他们的实验已经十分接近观察到 Efimov 共振现象, 然而, 由于实验样品温度是 1μK, 这个温度显然太 "热" 了, 以致不能清晰地观察到 Efimov 三原子共振.

理论预言 35 年后, 2005 年, 奥地利因斯布鲁克大学 Grimm 研究组终于在超冷铯原子气体中观察到了 Efimov 共振, 其在 10nK 的超冷温度下出现. 实际上该小组早在 2002 年就在 Cs 原子玻色–爱因斯坦凝聚中清晰地观察到该现象, 只是当时没有引起注意而已 [201]. 从该现象预言到实验实现整整用了 35 年, 巧合的是从 1935 年 Thomas 效应的提出到 Efimov 的预言, 也是 35 年. 更为欣喜的是, 十年后的 2015 年, 氦三体系的激发态终于被实验验证为一个 Efimov 态 [207].

目前看来, 实验上之所以实现 Efimov 共振如此艰辛, 是因为要形成 Efimov 三体态, 有两个条件是必要的, 即大的散射长度 a 和超低温度 [208]. 大的散射长度是相对于特征相互作用范围而言, 对于中性原子, 特征相互作用范围由 van der Waals 相互作用决定, 其典型值发现在 $30a_0$ (Li 原子)~$100a_0$ (Cs 原子), 其中 a_0 是玻尔半径. 要求有大的散射长度, 现在知道利用磁场调节的 Feshbach 共振就能达到这个目的. 在 Feshbach 共振中心附近有一个确定的范围存在 [也叫幺正范围, 见图 2.19(b)], 其中两通道相互作用问题得以简化, 在这个范围, 完全可以用散射长度这个参数来描述, 如弱束缚态在 $a > 0$ 时具有能量 $E_b = -\hbar^2/(ma^2)$. 之所以有超低温度的要求, 是因为超冷原子气体中碰撞能量极低 (典型值在 peV 范围), 相互作用由 s 波散射决定, 其他高分波影响可以忽略. 在这种情况下, s 波散射长度完全能够特征化两体相互作用.

为了更加清晰地显示 Efimov 共振发生时具备的条件, 再一次画出超冷碰撞时 Feshbach 共振的情形, 见图 2.19. 图 2.19(a) 为磁场调节 Feshbach 共振的两通道模型. 原子碰撞中 van der Waals 相互作用控制长程势并且用来界定长程物理和短程物理的边界. 超冷碰撞时 $E \to 0$, 原子态和束缚态 (上面虚线) 的能量差 δE 由不同的势支持, 能调节到共振. 开道中最后一个束缚态 (低虚线) 决定背景散射长度. 图 2.19(b) 显示了 s 波散射长度在共振处的发散行为. 图 2.19(c) 为束缚能, 子图为近 Feshbach 共振时的情形. 共振中心周围的阴影区域就是研究 Efimov 共振发生的最

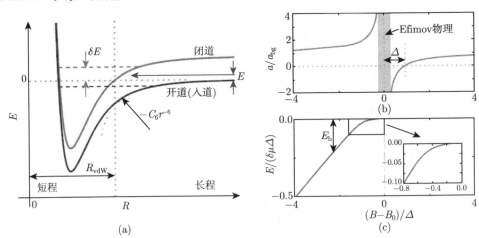

图 2.19 超冷碰撞时 Feshbach 共振 [209]

佳范围, 其宽度依赖于共振的特性. 原子在电子基态情况下, 长程势有一个由 van der Waals 相互作用的尾巴 $-C_6/r^6$, 常数 C_6 是碰撞原子的特征化参数, 其值根据第一性原理数值计算得到. van der Waals 相互作用的自然长度尺度由 van der Waals 尺度来表征, 其定义为 $R_\mathrm{vdW} = (mC_6/\hbar^2)^{1/4}/2$, 该参数也决定了原子碰撞中短程和长程相互作用的自然边界. 图 2.19(b) 中的阴影区域就是三体 Efimov 共振发生的区域[210].

Efimov 共振的实验观察激发了人们对该领域研究的极大热情, 后来人们陆续在不同超冷原子系统中观察到该现象, 理论和实验工作如雨后春笋般发展起来. 特别是在实验上, 人们不仅在超冷 ^{133}Cs 系统继续实现了 Efimov 共振, 并观察到了尺度率[211], 还在其他的玻色系统, 如 ^{39}K [212] 和 ^{7}Li [213,214], 以及玻色混合系统 ^{41}K^{87}Rb 中分别实现了 Efimov 三体态[215]. 同时, 在超冷费米系统 ^{6}Li 中也观察到了该现象[216-219]. 这些工作大多集中在 2009 年 (其中 2006 年 2 个, 2008 年 5 个, 2009 年 9 个), 发表在 Nature, Nature Physics 和 Phys. Rev. Lett. 等期刊上. 人们逐渐发现, 不仅能在超冷玻色系统和费米系统中实现 Efimov 三体态, 而且也能在许多其他的三粒子系统中实现. Efimov 本人对此感到非常兴奋, 并在文献 [208] 中描述到: "令人兴奋的是亲眼看见了这一量子力学奇迹的演化过程, 从值得怀疑的到病态的, 再到奇异的, 最后变成今天在超冷物理中的一个热门研究课题, 让童话变成了现实." (It has been heartening to witness the evolution of this miracle of quantum mechanics from questionable to pathological to exotic to being a hot topic of today's ultracold physics. The fairytale is becoming a reality.)

除此之外, 由于 Efimov 效应显示了在两体共振相互作用条件下普遍存在一系列三体束缚态, 人们自然地就想到, 是否在 $N-1$ 体共振相互作用下存在一系列 N 体束缚态? 最简单的推广就是三体 Efimov 共振情况下是否存在四体 Efimov 态? 该问题很早就引起了研究者的注意, 早在 Efimov 预言后不久的 1973 年, Amado 和 Greenwood[220] 就预言不会存在四体或更多体的 Efimov 态. 然而, 近年来 Hammer 等和 JILA 研究小组却预言每一个 Efimov 共振态都会伴随着两个四体态[221,222], 更为可喜的是 Innsbruck 研究组于 2009 年已经在超冷铯原子气体中实现了和 Efimov 三体共振相关的 Efimov 四体态[223]. 这一现象后来在四个全同玻色气体中得到进一步的发展[213,224-226]. 四体全同玻色子系统 Efimov 共振的基本图像如图 2.20 所示, 整个平面分成了四个部分, 分别对应于四个不同的态: 四个原子连续态、两体-原子-原子态、两体-两体态以及三体态、四体态. 实线表示在每一个 Efimov 三体态附近对应的一对四体态. Efimov 三体态导致出现了三体-原子阈值 (虚线). 图 2.20 中的两个箭头表示两个四体态形成的位置.

最近的理论研究已经将 Efimov 共振做了进一步的推进, 其中 $N = 5$、6、7 体或者更多体团簇态 (N 体 Borromean) 能被形成, 甚至在不存在束缚子系统时也毫不影响团簇态的形成 [225,227-229]. 文献 [230] 已经研究了 $N = 40$ 个原子的团簇态, 并计算了连续的 N 个玻色系统穿过对应的原子阈值时的散射长度的值. 目前, 人们称与 Efimov 效应相关的内容为 Efimov 物理 [231].

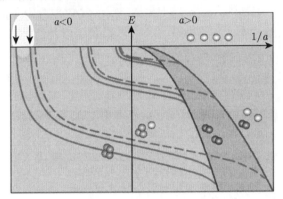

图 2.20　四体 Efimov 物理: 能量随散射长度的倒数的变化情况 [201]

　　Efimov 共振的实现, 就如 Feshbach 共振在研究两体量子力学相互作用的重要手段一样, 它将成为研究少体物理甚至多体物理的一个重要桥梁, 将又一次激起对超冷原子气体的研究热潮, 目前已初见端倪. 特别是, 所形成的 Efimov 分子态对超冷原子气体的稳定性有巨大影响, Efimov 态独有的分子构型不依赖于原子的种类, 因而, 首先可以用对所有原子都普适的方法进行合成, 然后作为中间态被转化到其他特定的分子态上, 形成超冷分子. 又由于 Efimov 态的多种构型, Efimov 共振自然而然也成为形成超冷多聚物分子的一个有力工具. 尽管这方面的研究仍处于起步探索阶段, 但作为一种可控分子合成的潜在方法, 人们对它未来的应用前景充满了无限期望.

　　形成三原子或者多原子分子是超冷原子分子领域的热点研究课题. 原则上得到多原子超冷分子是可能的, 要么直接冷却分子, 要么通过原子或者小分子缔合. 如果能将复杂得多聚物分子冷却到超低温度, 将是冷分子领域研究的一个重要进展, 并引起一个有趣的新方向. 例如, 可用于大分子碰撞截面的精密测量、分子光学以及控制超冷化学等.

　　一种潜在的形成冷多聚物分子的方法是缓冲气体制冷[131] 或者 Stark 减速 [132], 然而, 这些方法, 如 Stark 减速仍然不能达到超冷温度范围. 近几年, 实验上报道了一个直接冷却多原子分子的 Sisyphus 冷却 [232] 方法: 将氟代甲烷 (CH_3F) 分子冷却到 29 mK, 甚至温度还能低于 1mK. 另一个有希望产生超冷多原子分子的有力方法和磁场调节的 Feshbach 共振有关. 前面已经提到, 借助 Feshbach 共振和光缔合, 广义的受激拉曼绝热通道技术 [182] 首先被用于产生大量的深束缚态超冷双原子分子, 该方法利用一个啁啾耦合场去补偿粒子间相互作用的非线性效应.

目前,这种方法已经被用来在理论上形成同核和异核超冷三原子分子[185,186],这实际上是将 Feshbach 共振、光缔合以及受激拉曼绝热通道技术结合起来的一种技术路线.

Efimov 共振的实现无疑也给超冷分子的形成增添了活力,也许会成为又一个产生超冷多原子分子的有力工具. 通过借助磁场 Feshbach 共振和 Efimov 共振的特点,将玻色气体加载在光晶格上已经被提议用来产生三体 Efimov 分子[233]. 利用 Feshbach 共振形成的 Cs_2 分子和 Cs_4 分子态已经被观察到[234]. 借助三体 Efimov 共振和光缔合技术,广义的受激拉曼绝热通道技术已经用于形成同核和异核超冷四聚物分子[235,236]. 利用该技术,结合 Efimov 物理目前的发展,能否产生超冷多聚物分子,其稳定性和绝热性如何等均是值得研究的课题. 本书第 5 章就围绕这些问题展开,探讨多聚物分子的形成.

参 考 文 献

[1] SIBILLE A, PALMIER J F, LARUELLE F. Zener Interminiband resonant breakdown in superlattices[J]. Phys. Rev. Lett., 1998, 80: 4506-4509.

[2] IZMALKOV A, GRAJCAR M, IL'ICHEV E, et al. Observation of macroscopic Landau-Zener transitions in a superconducting device[J]. Europhys. Lett., 2004, 65(6): 844.

[3] BHARUCHA C F, MADISON K W, MORROW P R, et al. Observation of atomic tunneling from an accelerating optical potential[J]. Phys. Rev. A, 1997, 55: R857-R860.

[4] 刘杰. 玻色–爱因斯坦凝聚体动力学 —— 非线性隧穿、相干及不稳定性 [M]. 北京:科学出版社, 2009.

[5] 豆福全. 超冷原子系统中非线性隧穿及原子–分子转化的量子操控 [D]. 北京:北京理工大学, 2013.

[6] LANDAU L D. On the theory of transfer of energy at collisions II[J]. Phys. Z. Sowjetunion, 1932, 2: 46-51.

[7] ZENER C. Non-adiabatic crossing of energy levels[J]. Proc. R. Soc. London, 1932, 137: 696-702.

[8] STÜCKELBERG E C G. Theory of inelastic collisions between atoms (Theory of inelastic collisions between atoms, using two simultaneous differential equations)[J]. Helv. Phys. Acta, 1932, 5: 369-422.

[9] MAJORANA E. Atomi orientati in campo magnetico variabile[J]. Nuovo Cimento, 1932, 9: 43-50.

[10] GIACOMO F DI, NIKITIN E E. The Majorana formula and the Landau-Zener-Stückelberg treatment of the avoided crossing problem[J]. Physics Uspekhi, 2005, 48(5): 515-517.

[11] ROSEN N, ZENER C. Double Stern-Gerlach experiment and related collision phenomena[J]. Phys. Rev., 1932, 40(4): 502-507.

[12] SUOMINEN K A, GARRAWAY B. Population transfer in a level-crossing model with two time scales[J]. Phys. Rev. A, 1992, 45(1): 374.

[13] ALLEN L, EBERLY J H. Optical Resonance and Two-Level Atoms[M]. New York: Dover Publications, 1987.

[14] DEMKOV Y N, KUNIKE M. Hypergeometric models for the two-state approximation in collision theory[J]. Vestn. Leningr. Univ., 1969,16(3): 39.

[15] HIOE F, CARROLL C. Analytic solutions to the two-state problem for chirped pulses[J]. J. Opt. Soc. Am. B, 1985, 2: 497-502.

[16] KYOSEVA E S, VITANOV N V. Coherent pulsed excitation of degenerate multistate systems: Exact analytic solutions[J]. Phys. Rev. A, 2006, 73: 023420.

[17] BAMBINI A, BERMAN P R. Analytic solutions to the two-state problem for a class of coupling potentials[J]. Phys. Rev. A, 1981, 23: 2496-2501.

[18] NIKITIN E E. Resonance and non-resonance intermolecular energy exchange in molecular collisions[J]. Discuss. Faraday Soc., 1962, 33: 14-21.

[19] OSTROVSKY V. Nonstationary multistate Coulomb and multistate exponential models for nonadiabatic transitions[J]. Phys. Rev. A, 2003, 68(1): 012710.

[20] VITANOV N, STENHOLM S. Pulsed excitation of a transition to a decaying level[J]. Phys. Rev. A, 1997, 55(4): 2982.

[21] TOROSOV B, VITANOV N. Coherent control of a quantum transition by a phase jump[J]. Phys. Rev. A, 2007, 76(5): 053404.

[22] MACKIE M, KOWALSKI R, JAVANAINEN J. Bose-stimulated Raman adiabatic passage in photoassociation[J]. Phys. Rev. Lett., 2000, 84: 3803-3806.

[23] VASILEV G, VITANOV N. Coherent excitation of a two-state system by a linearly chirped Gaussian pulse[J]. J. Chem. Phys., 2005, 123: 174106.

[24] ROLAND J, CERF N J. Quantum search by local adiabatic evolution[J]. Phys. Rev. A, 2002, 65(4): 042308.

[25] WERNSDORFER W, SESSOLI R. Quantum phase interference and parity effects in magnetic molecular clusters[J]. Science, 1999, 284(5411): 133-135.

[26] FÖLDI P, BENEDICT M G, PEREIRA J M, et al. Dynamics of molecular nanomagnets in time-dependent external magnetic fields: Beyond the Landau-Zener-Stückelberg model[J]. Phys. Rev. B, 2007, 75: 104430.

[27] CHILD S. Molecular Collision Theory[M]. New York: Dover, 1974.

[28] SPREEUW R J C, VAN DRUTEN N J, BEIJERSBERGEN M W, et al. Classical realization of a strongly driven two-level system[J]. Phys. Rev. Lett., 1990, 65: 2642-2645.

[29] RUBBMARK J R, KASH M M, LITTMAN M G, et al. Dynamical effects at avoided level crossings: A study of the Landau-Zener effect using Rydberg atoms[J]. Phys. Rev. A, 1981, 23: 3107-3117.

[30] JONA-LASINIO M, MORSCH O, CRISTIANI M, et al. Asymmetric Landau-Zener tunneling in a periodic potential[J]. Phys. Rev. Lett., 2003, 91: 230406.

[31] WOO S J, CHOI S, BIGELOW N P. Controlling quasiparticle excitations in a trapped Bose-Einstein condensate[J]. Phys. Rev. A, 2005, 72: 021605.

参考文献

[32] MULLEN K, BEN-JACOB E, SCHUSS Z. Combined effect of Zener and quasiparticle transitions on the dynamics of mesoscopic Josephson junctions[J]. Phys. Rev. Lett., 1988, 60: 1097-1100.

[33] KHOMERIKI R, RUFFO S. Nonadiabatic Landau-Zener tunneling in waveguide arrays with a step in the refractive index[J]. Phys. Rev. Lett., 2005, 94: 113904.

[34] DREISOW F, SZAMEIT A, HEINRICH M, et al. Direct observation of Landau-Zener tunneling in a curved optical waveguide coupler[J]. Phys. Rev. A, 2009, 79: 055802.

[35] LIU J, LIU B, FU L B. Many-body effects on nonadiabatic Feshbach conversion in bosonic systems[J]. Phys. Rev. A, 2008, 78: 013618.

[36] LIU J, FU L B, LIU B, et al. Role of particle interactions in the Feshbach conversion of fermionic atoms to bosonic molecules[J]. New J. Phys., 2008, 10(12): 123018.

[37] VITANOV N V, GARRAWAY B M. Landau-Zener model: Effects of finite coupling duration[J]. Phys. Rev. A, 1996, 53: 4288-4304.

[38] ZHU C, NAKAMURA H. Semiclassical theory of multi-channel curve crossing problems: Landau-Zener case[J]. J. Chem. Phys., 1997, 106: 2599.

[39] ZHU C, NAKAMURA H. Semiclassical theory of multi-channel curve crossing problems: Nonadiabatic tunneling case[J]. J. Chem. Phys., 1997, 107: 7839.

[40] SHIMSHONI E, STERN A. Dephasing of interference in Landau-Zener transitions[J]. Phys. Rev. B, 1993, 47: 9523-9536.

[41] SAITO K, WUBS M, KOHLER S, et al. Dissipative Landau-Zener transitions of a qubit: Bath-specific and universal behavior[J]. Phys. Rev. B, 2007, 75: 214308.

[42] POKROVSKY V L, SUN D. Fast quantum noise in the Landau-Zener transition[J]. Phys. Rev. B, 2007, 76: 024310.

[43] CHEN Y A, HUBER S D, TROTZKY S, et al. Many-body Landau-Zener dynamics in coupled one-dimensional Bose liquids[J]. Nature Phys., 2010, 7(1): 61-67.

[44] WU B, NIU Q. Nonlinear Landau-Zener tunneling[J]. Phys. Rev. A, 2000, 61: 023402.

[45] ZOBAY O, GARRAWAY B M. Time-dependent tunneling of Bose-Einstein condensates[J]. Phys. Rev. A, 2000, 61: 033603.

[46] LIU J, FU L, OU B Y, et al. Theory of nonlinear Landau-Zener tunneling[J]. Phys. Rev. A, 2002, 66(2): 023404.

[47] LIU J, WU B, FU L, et al. Quantum step heights in hysteresis loops of molecular magnets[J]. Phys. Rev. B, 2002, 65(22): 224401.

[48] WU B, LIU J. Commutability between the semiclassical and adiabatic limits[J]. Phys. Rev. Lett., 2006, 96(2): 20405.

[49] SUQING D, FU L B, LIU J, et al. Effects of periodic modulation on the Landau-Zener transition[J]. Phys. Lett. A, 2005, 346(4): 315-320.

[50] LIU J, FU L B. Singularities of Berry connections inhibit the accuracy of the adiabatic approximation[J]. Phys. Lett. A, 2007, 370(1): 17-21.

[51] YANG L Y, FU L B, LIU J. Nonlinear Landau-Zener tunneling of Bose-Einstein condensate

in a spatially magnetic modulated trap[J]. Laser Phys., 2009, 19(4): 678-685.

[52] WANG G F, YE D F, FU L B, et al. Landau-Zener tunneling in a nonlinear three-level system[J]. Phys. Rev. A, 2006, 74(3): 033414.

[53] GARANIN D A, SCHILLING R. Effects of nonlinear sweep in the Landau-Zener-Stueckelberg effect[J]. Phys. Rev. B, 2002, 66: 174438.

[54] ALBIEZ M, GATI R, FÖLLING J, et al. Direct observation of tunneling and nonlinear self-trapping in a single bosonic Josephson junction[J]. Phys. Rev. Lett., 2005, 95: 010402.

[55] FU L B, LIU J. Quantum entanglement manifestation of transition to nonlinear self-trapping for Bose-Einstein condensates in a symmetric double well[J]. Phys. Rev. A, 2006, 74: 063614.

[56] LIU B, FU L B, YANG S P, et al. Josephson oscillation and transition to self-trapping for Bose-Einstein condensates in a triple-well trap[J]. Phys. Rev. A, 2007, 75(3): 033601.

[57] YE D F, FU L B, LIU J. Rosen-Zener transition in a nonlinear two-level system[J]. Phys. Rev. A, 2008, 77(1): 013402.

[58] LI S C, FU L B, DUAN W S, et al. Nonlinear Ramsey interferometry with Rosen-Zener pulses on a two-component Bose-Einstein condensate[J]. Phys. Rev. A, 2008, 78(6): 063621.

[59] FU L B, YE D F, LEE C, et al. Adiabatic Rosen-Zener interferometry with ultracold atoms[J]. Phys. Rev. A, 2009, 80(1): 013619.

[60] ISHKHANYAN A, CHERNIKOV G, NAKAMURA H. Rabi dynamics of coupled atomic and molecular Bose-Einstein condensates[J]. Phys. Rev. A, 2004, 70(5): 053611.

[61] ISHKHANYAN A, NAKAMURA H. Strong-coupling limit in cold-molecule formation via photoassociation or Feshbach resonance through Nikitin exponential resonance crossing[J]. Phys. Rev. A, 2006, 74(6): 063414.

[62] ISHKHANYAN A. Generalized formula for the Landau-Zener transition in interacting Bose-Einstein condensates[J]. Europhys. Lett., 2010, 90(3): 30007.

[63] NIELEN M, CHUANG I. Quantum Computation and Quantum Information[M]. Cambridge: Cambridge Univ. Press, 2000.

[64] 豆福全. 超冷原子系统中的高保真度量子操控 [D]. 北京: 北京应用物理与计算数学研究所, 2017.

[65] BERRY M. Transitionless quantum driving[J]. J. Phys. A: Mathematical and Theoretical, 2009, 42(36): 365303.

[66] LIM R, BERRY M. Superadiabatic tracking of quantum evolution[J]. J. Phys. A: Mathematical and General, 1991, 24(14): 3255.

[67] DEMIRPLAK M, RICE S A. Adiabatic population transfer with control fields[J]. The Journal of Physical Chemistry A, 2003, 107(46): 9937-9945.

[68] DEMIRPLAK M, RICE S A. On the consistency, extremal, and global properties of counter-diabatic fields[J]. The Journal of chemical physics, 2008, 129(15): 154111.

[69] CHEN X, RUSCHHAUPT A, SCHMIDT S, et al. Fast optimal frictionless atom cooling in harmonic traps: Shortcut to adiabaticity[J]. Phys. Rev. Lett., 2010, 104(6): 063002.

[70] IBÁÑEZ S, CHEN X, TORRONTEGUI E, et al. Multiple schrödinger pictures and dynamics in shortcuts to adiabaticity[J]. Phys. Rev. Lett., 2012, 109(10): 100403.

[71] CHEN X, LIZUAIN I, RUSCHHAUPT A, et al. Shortcut to adiabatic passage in two-and three-level atoms[J]. Phys. Rev. Lett., 2010, 105(12): 123003.

[72] TORRONTEGUI E, IBÁÑEZ S, MARTÍNEZ-GARAOT S, et al. Shortcuts to adiabaticity[J]. Adv. At. Mol. Opt. Phys, 2013, 62: 117-169.

[73] DEL CAMPO A. Shortcuts to adiabaticity by counterdiabatic driving[J]. Phys. Rev. Lett., 2013, 111(10): 100502.

[74] CAMPBELL S, DE CHIARA G, PATERNOSTRO M, et al. Shortcut to adiabaticity in the Lipkin-Meshkov-Glick model[J]. Phys. Rev. Lett., 2015, 114(17): 177206.

[75] LU M, XIA Y, SHEN L T, et al. Shortcuts to adiabatic passage for population transfer and maximum entanglement creation between two atoms in a cavity[J]. Phys. Rev. A, 2014, 89(1): 012326.

[76] TAKAHASHI K. Transitionless quantum driving for spin systems[J]. Phys. Rev. E, 2013, 87(6): 062117.

[77] PAUL K, SARMA A K. Shortcut to adiabatic passage in a waveguide coupler with a complex-hyperbolic-secant scheme[J]. Phys. Rev. A, 2015, 91(5): 053406.

[78] MASUDA S, GÜNGÖRDÜ U, CHEN X, et al. Fast control of topological vortex formation in Bose-Einstein condensates by counterdiabatic driving[J]. Phys. Rev. A, 2016, 93(1): 013626.

[79] SANTOS A C, SILVA R D, SARANDY M S. Shortcut to adiabatic gate teleportation[J]. Phys. Rev. A, 2016, 93(1): 012311.

[80] SUN Z, ZHOU L, XIAO G, et al. Finite-time Landau-Zener processes and counterdiabatic driving in open systems: Beyond Born, Markov, and rotating-wave approximations[J]. Phys. Rev. A, 2016, 93(1): 012121.

[81] LIANG Z T, YUE X, LV Q, et al. Proposal for implementing universal superadiabatic geometric quantum gates in nitrogen-vacancy centers[J]. Phys. Rev. A, 2016, 93(4): 040305.

[82] OKUYAMA M, TAKAHASHI K. From classical nonlinear integrable systems to quantum shortcuts to adiabaticity[J]. Phys. Rev. Lett., 2016, 117(7): 070401.

[83] IBÁÑEZ S, MARTÍNEZ-GARAOT S, CHEN X, et al. Shortcuts to adiabaticity for non-Hermitian systems[J]. Phys. Rev. A, 2011, 84(2): 023415.

[84] SCHAFF J F, SONG X L, VIGNOLO P, et al. Fast optimal transition between two equilibrium states[J]. Phys. Rev. A, 2010, 82(3): 033430.

[85] BASON M G, VITEAU M, MALOSSI N, et al. High-fidelity quantum driving[J]. Nat. Phys., 2012, 8: 147-152.

[86] MALOSSI N, BASON M G, VITEAU M, et al. Quantum driving protocols for a two-level system: From generalized Landau-Zener sweeps to transitionless control[J]. Phys. Rev. A, 2013, 87: 012116.

[87] ZHANG J, SHIM J H, NIEMEYER I, et al. Experimental implementation of assisted quantum adiabatic passage in a single spin[J]. Phys. Rev. Lett., 2013, 110(24): 240501.

[88] DU Y X, LIANG Z T, LI Y C, et al. Experimental realization of stimulated Raman shortcut-to-adiabatic passage with cold atoms[J]. Nature communications, 2016, 7: 12479.

[89] AN S, LV D, DEL CAMPO A, et al. Shortcuts to adiabaticity by counterdiabatic driving for trapped-ion displacement in phase space[J]. Nature communications, 2016, 7: 12999.

[90] ZHOU B B, BAKSIC A, RIBEIRO H, et al. Accelerated quantum control using superadiabatic dynamics in a solid-state lambda system[J]. Nature Physics, 2017, 13(4): 330-334.

[91] BAKSIC A, RIBEIRO H, CLERK A A. Speeding up adiabatic quantum state transfer by using dressed states[J]. Phys. Rev. Lett., 2016, 116(23): 230503.

[92] CHEN Y H, XIA Y, WU Q C, et al. Method for constructing shortcuts to adiabaticity by a substitute of counterdiabatic driving terms[J]. Phys. Rev. A, 2016, 93(5): 052109.

[93] SCHLOSSHAUER M. Decoherence, the measurement problem, and interpretations of quantum mechanics[J]. Rev. Mod. Phys., 2005, 76: 1267-1305.

[94] HOLLENBERG L C. Quantum control: through the quantum chicane[J]. Nature Phys., 2012, 8(2): 113-114.

[95] HEISENBERG W. ÜBer Den Anschaulichen Inhalt Der Quantentheoretischen Kinematik und Mechanik[M]//Original Scientific Papers Wissenschaftliche Originalarbeiten.[S.l.]. Berlin Hcidelberg: Springer, 1985.

[96] MANDELSTAM L, TAMM I. The Uncertainty Relation Between Energy and Time in Non-Relativistic Quantum Mechanics[M]//Selected Papers[S.l.]. Berlin Hcidelberg: Springer, 1991.

[97] EMMANOUILIDOU A, ZHAO X G, AO P, et al. Steering an eigenstate to a destination[J]. Phys. Rev. Lett., 2000, 85: 1626-1629.

[98] TADDEI M M, ESCHER B M, DAVIDOVICH L, et al. Quantum speed limit for physical processes[J]. Phys. Rev. Lett., 2013, 110: 050402.

[99] CHENEAU M, BARMETTLER P, POLETTI D, et al. Light-cone-like spreading of correlations in a quantum many-body system[J]. Nature, 2012, 481(7382): 484-487.

[100] DEL CAMPO A, EGUSQUIZA I L, PLENIO M B, et al. Quantum speed limits in open system dynamics[J]. Phys. Rev. Lett., 2013, 110: 050403.

[101] VILLAMIZAR D V, DUZZIONI E I. Quantum speed limit for a relativistic electron in a uniform magnetic field[J]. Phys. Rev. A, 2015, 92: 042106.

[102] HEGERFELDT G C. Driving at the quantum speed limit: Optimal control of a two-level system[J]. Phys. Rev. Lett., 2013, 111: 260501.

[103] CANEVA T, MURPHY M, CALARCO T, et al. Optimal control at the quantum speed limit[J]. Phys. Rev. Lett., 2009, 103: 240501.

[104] RONG X, GENG J, WANG Z, et al. Implementation of dynamically corrected gates on a single electron spin in diamond[J]. Phys. Rev. Lett., 2014, 112: 050503.

[105] RONG X, GENG J, SHI F, et al. Experimental fault-tolerant universal quantum gates with solid-state spins under ambient conditions[J]. Nature communications, 2015, 6: 8748.

[106] GENG J, WU Y, WANG X, et al. Experimental time-optimal universal control of spin qubits in solids[J]. Phys. Rev. Lett., 2016, 117: 170501.

[107] Synopsis: Time optimization in quantum computing[OL]. [2016-10-19]. Physics, https://physics.aps.org/synopsis-for/10.1103/PhysRevLett.117.170501/.

[108] TOROSOV B T, GUÉRIN S, VITANOV N V. High-fidelity adiabatic passage by composite sequences of chirped pulses[J]. Phys. Rev. Lett., 2011, 106: 233001.

[109] SCHRAFT D, HALFMANN T, GENOV G T, et al. Experimental demonstration of composite adiabatic passage[J]. Physical Review A, 2013, 88(6): 063406.

[110] 豆福全, 郑伟强. 两能级系统的高保真度布居数反转 [J]. 科学通报, 2016, 61(20): 2309-2315.

[111] DAEMS D, RUSCHHAUPT A, SUGNY D, et al. Robust quantum control by a single-shot shaped pulse[J]. Phys. Rev. Lett., 2013, 111(5): 050404.

[112] KÖHLER T, GÓRAL K, JULIENNE P S. Production of cold molecules via magnetically tunable Feshbach resonances[J]. Rev. Mod. Phys., 2006, 78: 1311-1361.

[113] JONES K M, TIESINGA E, LETT P D, et al. Ultracold photoassociation spectroscopy: Long-range molecules and atomic scattering[J]. Rev. Mod. Phys., 2006, 78: 483-535.

[114] GIORGINI S, PITAEVSKII L P, STRINGARI S. Theory of ultracold atomic Fermi gases[J]. Rev. Mod. Phys., 2008, 80: 1215-1274.

[115] DE MELO C A S. When fermions become bosons: Pairing in ultracold gases[J]. Physics Today, 2008, 61: 45.

[116] CARR L D, DEMILLE D, KREMS R V, et al. Cold and ultracold molecules: Science, technology and applications[J]. New J. Phys., 2009, 11(5): 055049.

[117] CHIN C, GRIMM R, JULIENNE P, et al. Feshbach resonances in ultracold gases[J]. Rev. Mod. Phys., 2010, 82: 1225-1286.

[118] FRIEDRICH B, DOYLE J M. Why are cold molecules so hot?[J]. Chem. Phys. Chem., 2009, 10(4): 604-623.

[119] WYNAR R, FREELAND R, HAN D, et al. Molecules in a Bose-Einstein condensate[J]. Science, 2000, 287(5455): 1016-1019.

[120] BARTENSTEIN M, ALTMEYER A, RIEDL S, et al. Precise determination of ^{6}Li cold collision parameters by radio-frequency spectroscopy on weakly bound molecules[J]. Phys. Rev. Lett., 2005, 94: 103201.

[121] DEMILLE D. Quantum computation with trapped polar molecules[J]. Phys. Rev. Lett., 2002, 88: 067901.

[122] REY A M. The super cool atom thermometer[J]. Physics, 2009, 2: 103.

[123] WELD D M, MEDLEY P, MIYAKE H, et al. Spin gradient thermometry for ultracold atoms in optical lattices[J]. Phys. Rev. Lett., 2009, 103: 245301.

[124] OSPELKAUS S, NI K K, WANG D, et al. Quantum-state controlled chemical reactions of ultracold potassium-rubidium molecules[J]. Science, 2010, 327(5967): 853-857.

[125] HUTSON J M. Ultracold chemistry[J]. Science, 2010, 327(5967): 788-789.

[126] JIN D S, YE J. Polar molecules in the quantum regime[J]. Physics Today, 2011, 64: 27.

[127] 翟荟. 极性分子量子气体 [J]. 物理, 2011, 40(5): 336-337.

[128] SANTOS L, SHLYAPNIKOV G V, ZOLLER P, et al. Bose-Einstein condensation in trapped dipolar gases[J]. Phys. Rev. Lett., 2000, 85: 1791-1794.

[129] KREMS R V. Ultracold controlled chemistry[J]. Physics, 2010, 3: 10.

[130] HUTSON J M, SOLDAN P. Molecule formation in ultracold atomic gases[J]. International Rev. Phys. Chem., 2006, 25(4): 497-526.

[131] WEINSTEIN J D, DECARVALHO R, GUILLET T, et al. Magnetic trapping of calcium monohydride molecules at millikelvin temperatures[J]. Nature, 1998, 395(6698): 148-150.

[132] FULTON R, BISHOP A I, BARKER P. Optical Stark decelerator for molecules[J]. Phys. Rev. Lett., 2004, 93(24): 243004.

[133] FRIEDRICH B. Slowing of supersonically cooled atoms and molecules by time-varying non-resonant induced dipole forces[J]. Phys. Rev. A, 2000, 61: 025403.

[134] FULTON R, BISHOP A, SHNEIDER M, et al. Controlling the motion of cold molecules with deep periodic optical potentials[J]. Nature Phys., 2006, 2(7): 465-468.

[135] ELIOFF M S, VALENTINI J J, CHANDLER D W. Subkelvin cooling no molecules via "billiard-like" collisions with argon[J]. Science, 2003, 302(5652): 1940-1943.

[136] LIU N N, LOESCH H. Kinematic slowing of molecules formed by reactive collisions[J]. Phys. Rev. Lett., 2007, 98: 103002.

[137] GUPTA M, HERSCHBACH D. A mechanical means to produce intense beams of slow molecules[J]. The Journal of Physical Chemistry A, 1999, 103(50): 10670-10673.

[138] NIKITIN E, DASHEVSKAYA E, ALNIS J, et al. Measurement and prediction of the speed-dependent throughput of a magnetic octupole velocity filter including nonadiabatic effects[J]. Phys. Rev. A, 2003, 68: 023403.

[139] SHUMAN E, BARRY J, DEMILLE D. Laser cooling of a diatomic molecule[J]. Nature, 2010, 467(7317): 820-823.

[140] FESHBACH H. Unified theory of nuclear reactions[J]. Ann. Phys., 1958, 5(4): 357-390.

[141] FESHBACH H. A unified theory of nuclear reactions. II[J]. Anna. Phys., 1962, 19(2): 287-313.

[142] TIESINGA E, VERHAAR B J, STOOF H T C. Threshold and resonance phenomena in ultracold ground-state collisions[J]. Phys. Rev. A, 1993, 47: 4114-4122.

[143] INOUYE S, ANDREWS M, STENGER J, et al. Observation of Feshbach resonance in a Bose-Einstein condensate[J]. Nature, 1998, 392: 151.

[144] CHIN C. The birth of ultracold molecules into the world of quantum gases[J]. AAPPS Bulletin, 2004, 14(5): 14-21.

[145] 金政. 超冷分子的诞生与分子玻色-爱因斯坦凝聚 [J]. 物理双月刊, 2005, 27(2): 403-411.

[146] 孟少英. 超冷原子-分子暗态的绝热性和稳定性 [D]. 绵阳: 中国工程物理研究院, 2009.

[147] DONLEY E A, CLAUSSEN N R, THOMPSON S T, et al. Atom-molecule coherence in a Bose-Einstein condensate[J]. Nature, 2002, 417(6888): 529-533.

[148] HERBIG J, KRAEMER T, MARK M, et al. Preparation of a pure molecular quantum gas[J]. Science, 2003, 301(5639): 1510-1513.

[149] DÜRR S, VOLZ T, MARTE A, et al. Observation of molecules produced from a Bose-Einstein condensate[J]. Phys. Rev. Lett., 2004, 92: 020406.

[150] XU K, MUKAIYAMA T, ABO-SHAEER J R, et al. Formation of quantum-degenerate sodium molecules[J]. Phys. Rev. Lett., 2003, 91: 210402.

[151] GREINER M, REGAL C A, JIN D S. Emergence of a molecular Bose-Einstein condensate from a Fermi gas[J]. Nature, 2003, 426(6966): 537-540.

[152] STRECKER K E, PARTRIDGE G B, HULET R G. Conversion of an atomic Fermi gas to a long-lived molecular Bose gas[J]. Phys. Rev. Lett., 2003, 91: 080406.

[153] CUBIZOLLES J, BOURDEL T, KOKKELMANS S J J M F, et al. Production of long-lived ultracold li_2 molecules from a Fermi gas[J]. Phys. Rev. Lett., 2003, 91: 240401.

[154] JOCHIM S, BARTENSTEIN M, ALTMEYER A, et al. Pure gas of optically trapped molecules created from fermionic atoms[J]. Phys. Rev. Lett., 2003, 91: 240402.

[155] INOUYE S, GOLDWIN J, OLSEN M L, et al. Observation of heteronuclear Feshbach Resonances in a mixture of bosons and fermions[J]. Phys. Rev. Lett., 2004, 93: 183201.

[156] FERLAINO F, D'ERRICO C, ROATI G, et al. Feshbach spectroscopy of a K-Rb atomic mixture[J]. Phys. Rev. A, 2006, 73: 040702.

[157] OLSEN M L, PERREAULT J D, CUMBY T D, et al. Coherent atom-molecule oscillations in a Bose-Fermi mixture[J]. Phys. Rev. A, 2009, 80: 030701.

[158] STAN C A, ZWIERLEIN M W, SCHUNCK C H, et al. Observation of Feshbach resonances between two different atomic species[J]. Phys. Rev. Lett., 2004, 93: 143001.

[159] VOIGT A C, TAGLIEBER M, COSTA L, et al. Ultracold heteronuclear Fermi-Fermi molecules[J]. Phys. Rev. Lett., 2009, 102: 020405.

[160] PAPP S B, WIEMAN C E. Observation of heteronuclear Feshbach molecules from a ^{85}Rb-^{87}Rb Gas[J]. Phys. Rev. Lett., 2006, 97: 180404.

[161] WEBER C, BARONTINI G, CATANI J, et al. Association of ultracold double-species bosonic molecules[J]. Phys. Rev. A, 2008, 78: 061601.

[162] LIU J, LIU B, FU L B. Many-body effects on nonadiabatic Feshbach conversion in bosonic systems[J]. Phys. Rev. A, 2008, 78(1): 013618.

[163] LI J, YE D F, MA C, et al. Role of particle interactions in a many-body model of Feshbach-molecule formation in bosonic systems[J]. Phys. Rev. A, 2009, 79(2): 025602.

[164] LIU B, FU L B, LIU J. Shapiro-like resonance in ultracold molecule production via an oscillating magnetic field[J]. Phys. Rev. A, 2010, 81(1): 013602.

[165] 孟少英, 刘杰. 超冷原子-分子转化动力学: 受激拉曼绝热过程 [J]. 物理学进展, 2010, 30(3): 280-295.

[166] THORSHEIM H R, WEINER J, JULIENNE P S. Laser-induced photoassociation of ultracold sodium atoms[J]. Phys. Rev. Lett., 1987, 58: 2420-2423.

[167] STWALLEY W C, WANG H. Photoassociation of ultracold atoms: A new spectroscopic technique[J]. J. Mol. Spectros., 1999, 195(2): 194-228.

[168] FIORETTI A, COMPARAT D, CRUBELLIER A, et al. Formation of cold Cs_2 molecules through photoassociation[J]. Phys. Rev. Lett., 1998, 80: 4402-4405.

[169] NIKOLOV A N, EYLER E E, WANG X T, et al. Observation of ultracold ground-state potassium molecules[J]. Phys. Rev. Lett., 1999, 82: 703-706.

[170] NIKOLOV A N, ENSHER J R, EYLER E E, et al. Efficient production of ground-state potassium molecules at sub-mk temperatures by two-step photoassociation[J]. Phys. Rev. Lett., 2000, 84: 246-249.

[171] KERMAN A J, SAGE J M, SAINIS S, et al. Production and state-selective detection of ultracold rbcs molecules[J]. Phys. Rev. Lett., 2004, 92: 153001.

[172] SAGE J M, SAINIS S, BERGEMAN T, et al. Optical production of ultracold polar molecules[J]. Phys. Rev. Lett., 2005, 94: 203001.

[173] WANG D, QI J, STONE M F, et al. Photoassociative production and trapping of ultracold KRb molecules[J]. Phys. Rev. Lett., 2004, 93: 243005.

[174] BERGMANN K, THEUER H, SHORE B W. Coherent population transfer among quantum states of atoms and molecules[J]. Rev. Mod. Phys., 1998, 70: 1003-1025.

[175] WINKLER K, THALHAMMER G, THEIS M, et al. Atom-molecule dark states in a Bose-Einstein condensate[J]. Phys. Rev. Lett., 2005, 95: 063202.

[176] WINKLER K, LANG F, THALHAMMER G, et al. Coherent optical transfer of feshbach molecules to a lower vibrational state[J]. Phys. Rev. Lett., 2007, 98: 043201.

[177] NI K K, OSPELKAUS S, DE MIRANDA M, et al. A high phase-space-density gas of polar molecules[J]. Science, 2008, 322(5899): 231-235.

[178] MACKIE M, HÄRKÖNEN K, COLLIN A, et al. Improved efficiency of stimulated Raman adiabatic passage in photoassociation of a Bose-Einstein condensate[J]. Phys. Rev. A, 2004, 70: 013614.

[179] MACKIE M, COLLIN A, JAVANAINEN J. Comment on "Stimulated Raman adiabatic passage from an atomic to a molecular Bose-Einstein condensate"[J]. Phys. Rev. A, 2005, 71: 017601.

[180] DRUMMOND P D, KHERUNTSYAN K V, HEINZEN D J, et al. Stimulated Raman adiabatic passage from an atomic to a molecular Bose-Einstein condensate[J]. Phys. Rev. A, 2002, 65: 063619.

[181] MACKIE M. Feshbach-stimulated photoproduction of a stable molecular Bose-Einstein condensate[J]. Phys. Rev. A, 2002, 66: 043613.

[182] LING H Y, PU H, SEAMAN B. Creating a stable molecular condensate using a generalized Raman adiabatic passage scheme[J]. Phys. Rev. Lett., 2004, 93: 250403.

[183] PU H, MAENNER P, ZHANG W, et al. Adiabatic condition for nonlinear systems[J]. Phys. Rev. Lett., 2007, 98: 050406.

[184] LING H Y, MAENNER P, ZHANG W, et al. Adiabatic theorem for a condensate system in an atom-molecule dark state[J]. Phys. Rev. A, 2007, 75: 033615.

[185] JING H, CHENG J, MEYSTRE P. Coherent atom-trimer conversion in a repulsive Bose-Einstein condensate[J]. Phys. Rev. Lett., 2007, 99: 133002.

[186] JING H, CHENG J, MEYSTRE P. Coherent generation of triatomic molecules from ultracold atoms[J]. Phys. Rev. A, 2008, 77: 043614.

[187] MENG S Y, FU L B, LIU J. Adiabatic fidelity for atom-molecule conversion in a nonlinear

three-level Λ system[J]. Phys. Rev. A, 2008, 78(5): 053410.

[188] MENG S Y, FU L B, CHEN J, et al. Linear instability and adiabatic fidelity for the dark state in a nonlinear atom-trimer conversion system[J]. Phys. Rev. A, 2009, 79(6): 063415.

[189] ENOMOTO K, KASA K, KITAGAWA M, et al. Optical Feshbach resonance using the intercombination transition[J]. Phys. Rev. Lett., 2008, 101: 203201.

[190] PELLEGRINI P, GACESA M, CÔTÉ R. Giant formation rates of ultracold molecules via Feshbach-optimized photoassociation[J]. Phys. Rev. Lett., 2008, 101: 053201.

[191] MACKIE M, FENTY M, SAVAGE D, et al. Cross-molecular coupling in combined photoassociation and Feshbach resonances[J]. Phys. Rev. Lett., 2008, 101: 040401.

[192] XU X Q, LU L H, LI Y Q. Enhancing molecular conversion efficiency by a magnetic field pulse sequence[J]. Phys. Rev. A, 2009, 80: 033621.

[193] SABBAH H, BIENNIER L, SIMS I R, et al. Understanding reactivity at very low temperatures: The reactions of oxygen atoms with alkenes[J]. Science, 2007, 317(5834): 102-105.

[194] LARSEN J J, WENDT-LARSEN I, STAPELFELDT H. Controlling the branching ratio of photodissociation using aligned molecules[J]. Phys. Rev. Lett., 1999, 83: 1123-1126.

[195] HORNBERGER K, UTTENTHALER S, BREZGER B, et al. Collisional decoherence observed in matter wave interferometry[J]. Phys. Rev. Lett., 2003, 90: 160401.

[196] FLAMBAUM V V, KOZLOV M G. Enhanced sensitivity to the time variation of the fine-structure constant and m_p/m_e in diatomic molecules[J]. Phys. Rev. Lett., 2007, 99: 150801.

[197] BARTELS R A, WEINACHT T C, WAGNER N, et al. Phase modulation of ultrashort light pulses using molecular rotational wave packets[J]. Phys. Rev. Lett., 2001, 88: 013903.

[198] EFIMOV V. Energy levels arising from resonant two-body forces in a three-body system[J]. Phys. Lett. B, 1970, 33(8): 563-564.

[199] KRAEMER T, MARK M, WALDBURGER P, et al. Evidence for Efimov quantum states in an ultracold gas of caesium atoms[J]. Nature, 2006, 440(7082): 315-318.

[200] THOMAS L H. The interaction between a neutron and a Proton and the structure of H^3[J]. Phys. Rev., 1935, 47: 903-909.

[201] FERLAINO F, GRIMM R. Forty years of Efimov physics: How a bizarre prediction turned into a hot topic[J]. Physics, 2010, 3(9): 102.

[202] AMADO R D, NOBLE J V. Efimov's effect: A new pathology of three-particle systems. II[J]. Phys. Rev. D, 1972, 5: 1992-2002.

[203] LIM T K, DUFFY S K, DAMER W C. Efimov state in the ^{4}He trimer[J]. Phys. Rev. Lett., 1977, 38: 341-343.

[204] BRÜHL R, KALININ A, KORNILOV O, et al. Matter wave diffraction from an inclined transmission grating: Searching for the elusive ^{4}He trimer Efimov state[J]. Phys. Rev. Lett., 2005, 95: 063002.

[205] BACCARELLI I, DELGADO-BARRIO G, GIANTURCO F, et al. Searching for Efimov states in triatomic systems: The case of $LiHe_2$[J]. Europhys. Lett., 2000, 50(5): 567-573.

[206] VULETIĆ V, KERMAN A J, CHIN C, et al. in Laser Spectroscopy, XIV International

Conference[C]. Austria. edited by Blatt R, Eschner J, Leibfried D, and Schmidt-Kaler F, World Scientific, Singapore, 1999.

[207] KUNITSKI M, ZELLER S, VOIGTSBERER J, et al. Observation of the Efimov state of the helium trimer[J]. Science, 2015, 348(6234): 551-555.

[208] EFIMOV V. Few-body physics: Giant trimers true to scale[J]. Nature Phys., 2009, 5(8): 533-534.

[209] CHIN C, GRIMM R, JULIENNE P, et al. Feshbach resonances in ultracold gases[J]. Rev. Mod. Phys., 2010, 82: 1225-1286.

[210] FERLAINO F, ZENESINI A, BERNINGER M, et al. Efimov resonances in ultracold quantum gases[J]. Few-Body Systems, 2011, 51: 113-133.

[211] KNOOP S, FERLAINO F, MARK M, et al. Observation of an Efimov-like trimer resonance in ultracold atom-dimer scattering[J]. Nature Phys., 2009, 5(3): 227-230.

[212] ZACCANTI M, DEISSLER B, D'ERRICO C, et al. Observation of an Efimov spectrum in an atomic system[J]. Nature Phys., 2009, 5(8): 586-591.

[213] POLLACK S E, DRIES D, HULET R G. Universality in three-and four-body bound states of ultracold atoms[J]. Science, 2009, 326(5960): 1683-1685.

[214] GROSS N, SHOTAN Z, KOKKELMANS S, et al. Observation of universality in ultracold ^{7}Li three-body recombination[J]. Phys. Rev. Lett., 2009, 103: 163202.

[215] BARONTINI G, WEBER C, RABATTI F, et al. Observation of heteronuclear atomic Efimov resonances[J]. Phys. Rev. Lett., 2009, 103: 043201.

[216] OTTENSTEIN T B, LOMPE T, KOHNEN M, et al. Collisional stability of a three-component degenerate Fermi gas[J]. Phys. Rev. Lett., 2008, 101: 203202.

[217] HUCKANS J H, WILLIAMS J R, HAZLETT E L, et al. Three-body recombination in a three-state Fermi gas with widely tunable interactions[J]. Phys. Rev. Lett., 2009, 102: 165302.

[218] WILLIAMS J R, HAZLETT E L, HUCKANS J H, et al. Evidence for an excited-state Efimov trimer in a three-component Fermi gas[J]. Phys. Rev. Lett., 2009, 103: 130404.

[219] WENZ A N, LOMPE T, OTTENSTEIN T B, et al. Universal trimer in a three-component Fermi gas[J]. Phys. Rev. A, 2009, 80: 040702.

[220] AMADO R D, GREENWOOD F C. There is no efimov effect for four or more particles[J]. Phys. Rev. D, 1973, 7: 2517-2519.

[221] HAMMER H W, PLATTER L. Universal properties of the four-body system with large scattering length[J]. Eur. Phys. J. A-Hadrons and Nuclei, 2007, 32(1): 113-120.

[222] VON STECHER J, D'INCAO J P, GREENE C H. Signatures of universal four-body phenomena and their relation to the Efimov effect[J]. Nature Phys., 2009, 5(6): 417-421.

[223] FERLAINO F, KNOOP S, BERNINGER M, et al. Evidence for universal four-body states tied to an Efimov trimer[J]. Phys. Rev. Lett., 2009, 102: 140401.

[224] SCHMIDT R, MOROZ S. Renormalization-group study of the four-body problem[J]. Phys. Rev. A, 2010, 81: 052709.

[225] YAMASHITA M T, FEDOROV D V, JENSEN A S. Universality of brunnian (N-body borromean) four- and five-body systems[J]. Phys. Rev. A, 2010, 81: 063607.

[226] HADIZADEH M R, YAMASHITA M T, TOMIO L, et al. Scaling properties of universal tetramers[J]. Phys. Rev. Lett., 2011, 107: 135304.

[227] THØGERSEN M, FEDOROV D V, JENSEN A S. N-body Efimov states of trapped bosons[J]. Europhys. Lett., 2008, 83(3): 30012.

[228] VON STECHER J. Five- and six-body resonances tied to an Efimov trimer[J]. Phys. Rev. Lett., 2011, 107: 200402.

[229] VON STECHER J. Weakly bound cluster states of Efimov character[J]. J. Phys. B: Atomic, Molecular and Optical Physics, 2010, 43(10): 101002.

[230] HANNA G J, BLUME D. Energetics and structural properties of three-dimensional bosonic clusters near threshold[J]. Phys. Rev. A, 2006, 74: 063604.

[231] BRAATEN E, HAMMER H W. Efimov physics in cold atoms[J]. Ann. Phys., 2007, 322(1): 120-163.

[232] ZEPPENFELD M, ENGLERT B G E, GLÖCKNER R, et al. Sisyphus cooling of electrically trapped polyatomic molecules[J]. Nature, 2012, 491: 570.

[233] STOLL M, KÖHLER T. Production of three-body Efimov molecules in an optical lattice[J]. Phys. Rev. A, 2005, 72: 022714.

[234] CHIN C, KRAEMER T, MARK M, et al. Observation of Feshbach-like resonances in collisions between ultracold molecules[J]. Phys. Rev. Lett., 2005, 94: 123201.

[235] JING H, JIANG Y. Coherent atom-tetramer conversion: Bright-state versus dark-state schemes[J]. Phys. Rev. A, 2008, 77: 065601.

[236] LI G Q, PENG P. Formation of a heteronuclear tetramer A_3B via Efimov-resonance-assisted stimulated Raman adiabatic passage[J]. Phys. Rev. A, 2011, 83: 043605.

第 3 章　超冷原子系统中的非线性隧穿动力学

非线性隧穿动力学是超冷原子系统量子操控中的一个重要研究内容. 本章介绍非线性隧穿动力学的两个内容: 具有粒子间相互作用和非线性扫描两种非线性的两能级系统的 Landau-Zener 隧穿和具有粒子间相互作用的非线性 Demkov-Kunike 跃迁. 在 Landau-Zener 隧穿中, 由于两种非线性的相互竞争, 在不同参数范围内发生了一系列有趣的现象, 如绝热性的破坏、干涉现象的消失以及隧穿概率的不对称性. 对于 Demkov-Kunike 跃迁, 讨论了粒子间相互作用及其他控制场参数对跃迁动力学的影响, 发现了跃迁不对称等一些重要的规律.

3.1　广义 Landau-Zener 隧穿

作为量子力学中的一个基本的动力学过程, Landau-Zener 隧穿 (Landau-Zener tunneling, LZT) 早在 1932 年就分别由 Landau[1]、Zener[2]、Stückelberg[3] 和 Majorana[4] 独立提出. 该模型描述了一个哈密顿量依赖于时间的两能级量子系统, 两态之间的能级差随时间线性变化. 由于该模型的普遍性, 它已广泛应用于许多领域, 如分子纳米磁子 [5,6]、分子碰撞 [7]、量子点阵 [8]、Rydberg 原子 [9]、冷原子和加速光晶格中的玻色-爱因斯坦凝聚 [10,11]、冷 (超冷) 分子 [12,13]、场驱动的超晶格 [14]、流驱动的 Josephson 结 [15] 以及耦合波导管中的光波等 [16].

在过去许多年里, 最初的 Landau-Zener 隧穿已经从许多层面做了扩展, 包括不同的时间相互作用效应 [17]、非线性 [18–20]、有限耦合持续时间效应 [21]、多态动力学 [22,23]、退相干、噪声和耗散系统 [24] 以及多体量子动力学系统 [25] 等. 其中, 在非线性方面的扩展是最有趣的研究课题之一, 近年来引起了众多研究者的注意. 这方面的扩展主要体现在两个基本方面. 第一种情况是在描述两能级系统的线性方程中增加非线性项, 即相互作用项 (立方项) [18,19], 该模型描述了平均场近似下一个依赖于时间的两态系统的玻色-爱因斯坦凝聚. 该非线性 Landau-Zener 隧穿不仅在理论上有重要的意义, 而且在如纳米磁子的自旋隧穿 [26]、双势阱或者光晶格中的玻色-爱因斯坦凝聚 [18,27,28] 以及耦合波导管 [29] 中有着重要的应用. 后来该模型已被实验所证实 [10,25], 人们也解析地获得了一个广义的 Landau-Zener 隧穿公式 [30]. 第二种情况是在一些物理系统中能级随外场进行非线性扫描, 如在分子磁子 [31]、啁啾激光脉冲调制两态原子的相干激发 [32]、频率调制激光场的两态原子系统的相互作用 [33] 以及冷原子碰撞 [32] 等物理系统中. 这种非线性能级扫描方式

3.1 广义 Landau-Zener 隧穿

可用于操控量子系统在一个设想的态. 在单粒子隧穿系统中, 人们已经研究了一系列不同类型的非线性扫描方式 [34]. 这些研究启发人们: 如果在一个量子系统中同时存在粒子间相互作用和能级的非线性扫描, 会发生什么现象呢? 事实上, 相互作用的多体系统在平均场下的动力学本质上是非线性的, 实验上实现的线性扫描也仅仅是一种近似的结果.

本节主要聚焦于包含粒子间相互作用以及能级非线性扫描两种非线性对 Landau-Zener 隧穿过程的影响. 由于系统中存在两种非线性, 该 Landau-Zener 隧穿过程出现了许多不同于以前 Landau-Zener 隧穿的特殊性质 [35]. 在绝热极限下, 当粒子间相互作用超过一个临界值时, 绝热性会遭到破坏, 此时的隧穿概率基本不依赖于非线性能级扫描. 更为有趣的是, 在低速扫描时, 由非线性扫描引起的干涉现象会随着非线性粒子间相互作用的增强而逐渐消失. 在快速扫描范围, 隧穿概率主要依赖于非线性扫描, 并随着非线性扫描的增加而增大. 在中间扫描范围, 发现随着非线性扫描的增加, 会出现隧穿概率的不对称性. 为了进一步解释这些行为, 分析了系统的绝热能级并得到了稳相近似下的解析结果. 最后, 讨论了模型的一些可能应用.

3.1.1 粒子间相互作用和非线性扫描对 LZT 的影响

1. 模型

本节考虑的非线性两模模型由下述无量纲化的 Schrödinger 方程来描述:

$$i\frac{\partial}{\partial t}\begin{pmatrix} a \\ b \end{pmatrix} = H(t)\begin{pmatrix} a \\ b \end{pmatrix}, \tag{3.1}$$

哈密顿量为

$$H(t) = \frac{v}{2}\hat{\sigma}_x + \left[\frac{\gamma}{2} + \frac{c}{2}(|b|^2 - |a|^2)\right]\hat{\sigma}_z, \tag{3.2}$$

此即为方程 (2.6). 其中, a 和 b 是概率幅; $\hat{\sigma}_x$ 和 $\hat{\sigma}_z$ 是泡利矩阵; γ 和 v 分别是两模间的能级差和耦合强度; c 是描述粒子间相互作用的非线性参数, 不同于方程 (2.6) 的是, 这里 c 取正值时代表粒子间相互作用是相互吸引的 [18]. 总概率 $|a|^2 + |b|^2 = 1$ 守恒. 这里考虑耦合强度是常数, 能级差是关于时间的非线性函数:

$$v = \text{const}, \quad \gamma(t) = \text{sgn}(t)|\alpha t|^\beta, \tag{3.3}$$

其中, 参数 α 代表扫描速率; β 代表能级非线性扫描的控制参数.

当 $c = 0, \beta = 1$ 时, 对应于以前传统的 Landau-Zener 模型, 它能够精确求解. 假定系统初始态在 $a(-\infty) = 1, b(-\infty) = 0$, 非绝热基下在 $+\infty$ 时的隧穿概

率是 $|b(+\infty)|^2$. 绝热基下 (非绝热隧穿概率) 等于在非绝热基下无隧穿时的概率, $P = |a(+\infty)|^2$, 在本章中考虑的隧穿概率就是这个定义. 对 Landau-Zener 模型:

$$P = \exp\left(-\frac{\pi v^2}{2\alpha}\right). \tag{3.4}$$

很明显, 在绝热极限下隧穿概率是零. 这只是量子绝热定理的一种特殊情况: 当参数绝热改变时, 系统初始在本征态时将一直待在同样的瞬时本征态上.

当 $c \neq 0, \beta = 1$ 时, 对应于能级差随时间线性变化的非线性 Landau-Zener 模型. 如果相互作用 c 超过一个临界值, 绝热性会遭到破坏, 导致即使在绝热极限下也存在非零的隧穿概率. 结合经典绝热动力学和量子几何相位, 非线性量子系统中的非线性绝热定理已经被发现 [36].

当 $c = 0, \beta \neq 1$ 时, 对应于非线性扫描的 Landau-Zener 模型. 这种情况下的隧穿概率和扫描函数不同种类的奇性相关, 特别是在慢扫描时会出现干涉现象.

本节对同时具有粒子间相互作用和非线性扫描的 Landau-Zener 模型进行考察, 即当能级差随时间, 如 $\gamma(t) = \text{sgn}(t)|\alpha t|^\beta$ 变化时, 系统将如何演化. 事实上, Landau-Zener 模型、非线性 Landau-Zener 模型以及具有非线性能级扫描的 Landau-Zener 模型仅仅是这里所讨论的模型特例. 取耦合参数 $v = 1$. 因此, 弱相互作用、强相互作用以及慢扫描、快扫描分别指 $c \ll 1, c \gg 1$ 和 $\alpha \ll 1, \alpha \gg 1$.

2. 数值结果

由于出现了非线性, 系统的隧穿动力学发生了显著变化. 此时, Schrödinger 方程 (3.1) 不再能够解析求解. 采用 4-5 阶 Runge-Kutta 法跟踪系统的量子演化并计算其隧穿概率.

图 3.1 表示最终的隧穿概率在扫描速率从绝热极限到快速扫描范围变化时随粒子间相互作用 c 和能级扫描参数 β 的变化情况.

该图展现了粒子间相互作用和非线性能级扫描共同作用下 Landau-Zener 隧穿的基本特征.

(1) 绝热极限下, 隧穿概率不依赖于能级扫描参数 β 而主要依赖于粒子间相互作用 c, 且当 $c \leqslant v$ 时它的值近似为零. 然而, 当相互作用超过临界值时绝热性被破坏了, 导致出现一个非零的隧穿概率 [图 3.1(a)]. 在快速扫描范围, 隧穿概率随能级扫描参数 β 的增加而增大, 并且其值几乎不依赖于粒子间相互作用 [图 3.1(d)].

(2) 在低速扫描率时, 在弱相互作用 c 的情况下, 等高图出现了 "舌" 状结构, 并随能级扫描参数 β 而增大 [图 3.1(b)], 意味着出现了干涉现象. 隧穿概率的这种振荡在相互作用取某些值时变为零. 更为有趣的是, 这种隧穿概率的干涉现象会随着相互作用的增强而逐渐消失. 为了进一步了解隧穿概率的这种行为, 取参数 $\alpha = 0.1$, 并直接数值求解方程 (3.1), 获得了隧穿概率, 该结果如图 3.2(a) 所示.

3.1 广义 Landau-Zener 隧穿

(3) 在绝热极限和快速扫描之外的中间扫描速率范围, 隧穿概率和相互作用 c 与能级扫描参数 β 均有关, 并且随能级扫描参数 β 的增加具有不对称性. 在弱相互作用范围, 隧穿概率随能级扫描参数 β 增加而减小. 相反, 在强相互作用范围, 隧穿概率随能级扫描参数 β 的增加而增大 [图 3.1(b) 和 (c)]. 图 3.2(b) 中也显示了不同粒子间相互作用 c 情况下隧穿概率随能级扫描参数 β 的变化.

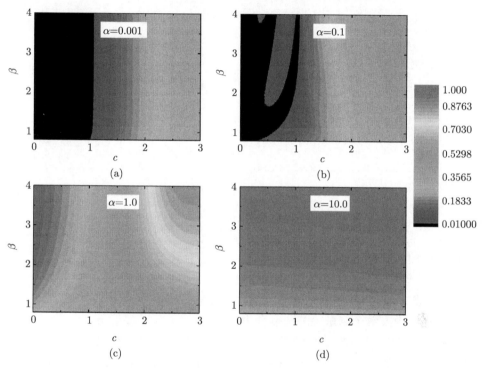

图 3.1 隧穿概率随相互作用 c 和能级扫描参数 β 在不同扫描速率 α 时的等高图

3. 理论分析

为了理解该隧穿动力学的物理机制, 本节通过计算绝热能级和强相互作用下稳相近似法所得到的隧穿概率解析表达式, 进行一些理论解释.

依赖于时间的非线性 Schrödinger 方程的非线性本征态和本征值定义如下:

$$H(t)\begin{pmatrix} a \\ b \end{pmatrix} = \epsilon(t)\begin{pmatrix} a \\ b \end{pmatrix} \tag{3.5}$$

也被称作绝热能级[37], 其中 $\epsilon(t)$ 为化学势.

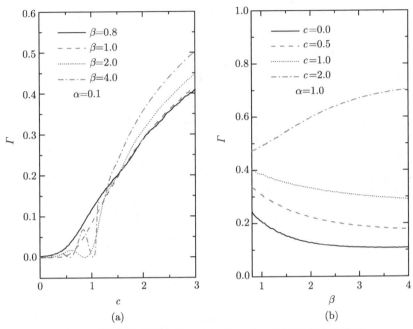

图 3.2 隧穿概率随参数 c (a) 和 β (b) 变化时的数值结果

能级将如何随粒子间相互作用 c 和能级扫描参数 β 变化而变化呢? 选取了几个典型参数, 计算结果如图 3.3 所示. 从图中 $v = 1.0$ 可以看到, 在相互作用 c 取相同值, 而参数 β 不同时, 能级在 $\alpha t = 0$ 时相互接触, 在 $\alpha t = \pm 1$ 时相交. 能级拓扑结构的变化将极大地影响系统的隧穿动力学.

在强非线性扫描控制参数情况下, 即 $\beta > 1$ 时, 低能级在弱相互作用时, 在 $\alpha t = 0$ 周围出现一个平台, 导致在该范围有一个宽的共振窗口, 以致系统在低能级和高能级之间振荡. 然而, 随着相互作用的逐渐加强, 低能级的平台变得越来越窄, 直到 $c > v$ 时一个环状结构出现, 该环的范围为 $-\gamma_c \leqslant \alpha t \leqslant \gamma_c$, 其中 $\gamma_c = (c^{2/3} - v^{2/3})^{3/2\beta}$ (出现环状结构也可以理解为一个宏观自俘获现象[25]). 其间, 共振范围也变得越来越窄甚至消失. 因此, 在弱相互作用范围会出现干涉现象, 而在强相互作用范围, 干涉现象会逐渐消失. 与此同时, 环状结构也使绝热性遭到破坏, 导致在绝热极限下也会出现一个从下能级到上能级的非零概率隧穿.

干涉现象的出现也能理解如下: 对弱相互作用范围, 隧穿概率的振荡由 $\sqrt{\gamma^2 + v_{\text{eff}}^2}$ (其中 v_{eff} 为系统的等效耦合强度[38]) 在 t 的上半复平面的奇性 (也称作隧穿点[32]) 决定. 当 $\beta > 1$ 时, 在 t 的上半复平面有两个接近于实轴且和虚轴对称的奇点, 它们的贡献相互干扰会导致隧穿概率的振荡[32,34]. 因此, 对非线性能级扫描, 在弱相互作用范围出现了隧穿概率的干涉现象. 在某些相互作用下, 隧穿概率为零则是由于两个奇点的作用相互抵消的结果.

3.1 广义 Landau-Zener 隧穿

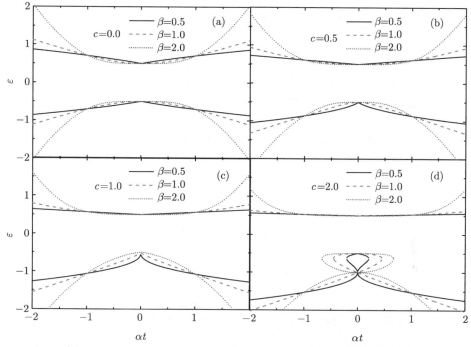

图 3.3 $c = 0.0$、0.5、1.0、2.0 和 $\beta = 0.5$、1.0、2.0 时的绝热能级

在弱相互作用情况下, 能级在小 β 值时变尖锐, 而在大 β 值时变得扁平. 对于一个固定的扫描速率, 能级的扁平将使得系统在 $\alpha t = 0$ 中心范围的持续时间加长, 抑制了隧穿概率, 导致了隧穿概率的降低[32] (需要注意的是隧穿概率的振荡在快速扫描时不存在). 另外, 尖锐的能级使得持续时间变短, 系统绝热跟随会更困难, 从而使隧穿概率增大. 在强相互作用情况下, 如文献 [25] 在耦合一维玻色流中情况一样, 随着 β 的增加, 隧穿概率会增大. 这是由于大 β 值导致了大的能级环状结构. 因而, 在中间扫描速率时, 在不同相互作用下, 隧穿概率随 β 的增加呈现了不对称性.

接下来, 为了进一步理解强相互作用下的隧穿动力学, 运用稳相近似[39,40] 得到了隧穿概率的解析表达式. 这里考虑强相互作用情形, 在这种情况下, 到上能级的隧穿概率即使在绝热极限下也接近于 1. 这样, 可以认为 Schrödinger 方程 (3.1) 中的 b 很小, 并且在整个过程中 $a \sim 1$, 可用摄动法来处理.

首先做变量代换:

$$a = a' \exp\left\{-\mathrm{i}\int_{-\infty}^{t}\left[\frac{\gamma}{2} + \frac{c}{2}(|b|^2 - |a|^2)\right]\mathrm{d}t\right\}, \tag{3.6}$$

$$b = b' \exp\left\{\mathrm{i}\int_{-\infty}^{t}\left[\frac{\gamma}{2} + \frac{c}{2}(|b|^2 - |a|^2)\right]\mathrm{d}t\right\}. \tag{3.7}$$

哈密顿量的对角项将消失, 有

$$b' = \frac{v}{2\mathrm{i}} \int_{-\infty}^{t} \mathrm{d}t \exp\left\{-\mathrm{i}\int_{-\infty}^{t}\left[\gamma + c(|b|^2 - |a|^2)\right] \mathrm{d}t\right\}. \tag{3.8}$$

需要自洽的对式 (3.8) 计算积分. 由于 c 非常大, 被积函数指数上的非线性项给出一个快速振荡相位, 对积分贡献很小. 积分的主要贡献来自稳相点 t_0 区域, 于是

$$|b|^2 = \left(\frac{v}{2}\right)^2 \left|\int_{-\infty}^{t} \exp\left[-\frac{\mathrm{i}}{2}\bar{\alpha}(t-t_0)^2\right] \mathrm{d}t\right|^2, \tag{3.9}$$

其中,

$$\bar{\alpha} = \left(\frac{\mathrm{d}\gamma}{\mathrm{d}t} + 2c\frac{\mathrm{d}|b|^2}{\mathrm{d}t}\right)_{t_0}, \tag{3.10}$$

稳相点 t_0 满足:

$$\gamma + c(2|b|^2 - 1) = 0. \tag{3.11}$$

对上述表达式做微分, 然后计算它在 t_0 点的值, 得到 $[(\mathrm{d}/\mathrm{d}t)|b|^2]_{t_0} = (v/2)^2\sqrt{2\pi/\bar{\alpha}}$①. 计算过程中用到了标准的 Fresnel 积分, 结合关系式 (3.10), 并且将 $\gamma = \mathrm{sgn}(t)|\alpha t|^\beta$ 代入方程 (3.10) 和方程 (3.11), 最终得到关于 $\bar{\alpha}$ 的一个封闭方程:

$$\bar{\alpha} = \alpha\beta c^{\frac{\beta-1}{\beta}} + 2c\left(\frac{v}{2}\right)^2\sqrt{\frac{2\pi}{\bar{\alpha}}}. \tag{3.12}$$

非绝热隧穿概率 Γ 为

$$\Gamma = 1 - |b|^2_{+\infty} = 1 - \left(\frac{v}{2}\right)^2 \left|\int_{-\infty}^{+\infty} \exp\left[-\frac{\mathrm{i}}{2}\bar{\alpha}(t-t_0)^2\right]\mathrm{d}t\right|^2$$
$$= 1 - \frac{\pi v^2}{2\bar{\alpha}}. \tag{3.13}$$

然后得到关于 Γ 的封闭方程:

$$\frac{1}{1-\Gamma} = \frac{1}{P} + \frac{2c}{\pi v}\sqrt{1-\Gamma}, \tag{3.14}$$

其中, $P = \pi v^2/2\alpha\beta c^{(\beta-1)/\beta}$. 在绝热极限下, 即 $1/P = 0$, 可以得到 $\Gamma = 1 - (\pi v/2c)^{2/3}$. 这表明在绝热极限下隧穿概率是非零的, 并且该概率不依赖于参数 β. 上面的解析结果方程 (3.14) 也已经通过直接数值求解 Schrödinger 方程 (3.1) 进行验证, 二者所得结果基本一致, 如图 3.4 所示. 从图中可以进一步验证, 强相互作用下, 在低速扫描时隧穿概率的干涉现象是不存在的 [图 3.4(a)]. 同时, 隧穿概率在中间扫描速率时随 β 增大而增大 [图 3.4(b)].

---
① 注意一些文献中的计算缺了一个 $\sqrt{2}$.

3.1 广义 Landau-Zener 隧穿

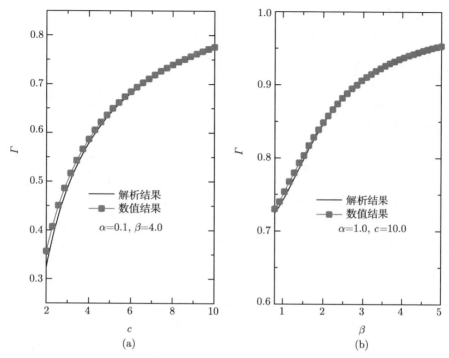

图 3.4 强相互作用时稳相近似的解析公式和直接数值求解 Schrödinger 方程 (3.1) 所得结果的对比

值得注意的是, 稳相近似也完全可以很好地处理快速扫描的情况, 这里略去有关的计算结果.

3.1.2 应用举例

上述模型可以直接应用于光晶格中两个 Bloch 带之间的玻色-爱因斯坦凝聚中, 其中无量纲化的 Schrödinger 方程包含了立方动能项 [参考文献 [18] 中的方程 (9)], 换句话说, 此时的能级扫描本身就是关于时间的非线性函数. 因此, 可以使用本节建立的非线性两能级模型作为理解光晶格中玻色-爱因斯坦凝聚的一个基本模型.

另一个有望用于观察这种非线性 Landau-Zener 隧穿行为的物理系统是耦合一维玻色流中的玻色-爱因斯坦凝聚[25]. 在这样一个系统中, 粒子间相互作用能通过沿着管方向的附加晶格控制, 而能级扫描的非线性可通过调节超晶格的相对相位来控制.

另外, 该非线性 Landau-Zener 模型还可以应用于许多其他物理系统中, 如在分子磁子[31] 和通过啁啾激光脉冲调控的两态原子系统的相干激发[32] 等. 在这样的物理系统中, 能级扫描本身就是时间的非线性函数. 因此, 只要考虑粒子间相互作用, 该理论就可以在这些系统中得以验证.

总之, 本节从数值和解析两方面研究了具有粒子间相互作用和非线性能级扫描的两能级 Landau-Zener 隧穿问题, 考察了粒子间相互作用和非线性能级扫描对隧穿动力学的共同影响. 研究表明, 这些非线性将极大地影响系统的隧穿动力学, 导致一系列有趣现象的发生. 另外, 也提议了一些有望在实验上观察到这种隧穿动力学的实际物理系统.

3.2 非线性 Demkov-Kunike 跃迁

量子跃迁在量子力学中是基本的现象之一, 在很多领域有重要的应用, 涉及化学反应、原子核系统、超晶格、超导器件和超冷原子系统[41-43] 等. 在线性两能级系统中, 已经有一些能精确求解的模型描述不同的跃迁过程, 包括 Landau-Zener-Stückelberg-Majorana 模型 [1-4]、Rosen-Zener 模型 [44]、Allen-Eberly 模型 [45,46]、Bambini-Berman 模型 [47] 等. 有一个更为广义的模型, 在一定条件下它可以简化为 Rosen-Zener 模型、Allen-Eberly 模型、Bambini-Berman 模型和 Landau-Zener 模型 [48,49], 该模型就是著名的 Demkov-Kunike 模型 [50,51].

自 Demkov 和 Kunike 第一次提出该模型后, 人们对其进行了广泛的研究, 并被应用到很多领域, 包括核磁共振 [52]、化学反应 [52]、激光冷却、原子光学 [53]、干涉 [54] 以及量子跃迁的相干控制 [55,56] 和波导管中 [57].

然而, 从严格意义上讲, 真实的物理系统是非线性的. 例如, 存在粒子间相互作用的两模玻色–爱因斯坦凝聚系统中就存在由平均场处理引起的粒子间相互作用. 近年来, Landau-Zener 模型和 Rosen-Zener 模型已经在非线性系统中被广泛研究 [18,35,36,58-64]. 在这些系统中, 不论理论还是实验, 人们对跃迁不对称性作了深入研究, 包括玻色–爱因斯坦凝聚系统 [65,66] 以及一维玻色流 [25] 等. 该性质对于理解系统的非线性动力学性质以及精确控制光晶格中的物质波等方面有重要意义 [67]. 同样, Demkov-Kunike 模型在非线性系统中也有初步研究, 如超冷原子的原子–分子转化 [68-70]、高保真复合绝热通道技术 [71] 等. 但是在存在粒子间相互作用的非线性系统中, Demkov-Kunike 跃迁动力学如何, 仍然不完全清楚 [72].

本节研究非线性两能级系统中的 Demkov-Kunike 跃迁, 主要关注粒子间相互作用对跃迁动力学的影响.

3.2.1 模型

非线性两种模型 Demkov-Kunike 跃迁同样由 Schrödinger 方程 (3.1) 描述, 只是在 Landau-Zener 模型中, 耦合强度为定值, 能级差随时间变化, 在 Rosen-Zener 模型中, 能级差为定值, 耦合强度则是由时间相关的外场调制. 不同于这两种模型,

3.2 非线性 Demkov-Kunike 跃迁

在 Demkov-Kunike 模型中,能级差和耦合强度都是由时间相关的外场调制,外场的形式为 [50,51]

$$\gamma(t) = \gamma_1 + \gamma_0 \tanh\left(\frac{t}{T}\right), \quad v(t) = v_0 \text{sech}\left(\frac{t}{T}\right), \quad (3.15)$$

其中,T 是外场的扫描周期,也叫脉冲宽度;γ_1 是静态失谐参数;γ_0 是啁啾参数;v_0 是最大耦合参数.

在下面的讨论中,假设初始的系统被完全制备在其中一个模态上,跃迁概率被定义为粒子布居在另一个能级上的概率. 换句话说,如果系统最初在模态 $\psi_1 = (a(-\infty), b(-\infty)) = (1,0)$,跃迁概率将是 $P = |b(+\infty)|^2$. 同样,当系统最初在模态 $\psi_2 = (a(-\infty), b(-\infty)) = (0,1)$,跃迁概率将是 $P = |a(+\infty)|^2$. 在数值计算中,时间 t 取足够大.

当 $c = 0$ 时,对应于线性系统中标准的 Demkov-Kunike 模型,它能够精确求解 [49]:

$$P = \frac{\cosh(\pi T \gamma_0) - \cosh[\pi T (\gamma_0^2 - v_0^2)^{\frac{1}{2}}]}{\cosh(\pi T \gamma_1) + \cosh(\pi T \gamma_0)}. \quad (3.16)$$

当 $c = 0, \gamma_1 = 0$ 时,对应于 Allen-Eberly 模型 [49]:

$$P = 1 - \frac{\cosh^2\left[\frac{\pi T (\gamma_0^2 - v_0^2)^{\frac{1}{2}}}{2}\right]}{\cosh^2\left(\frac{\pi T \gamma_0}{2}\right)}. \quad (3.17)$$

当 $c = 0, \gamma_0 = 0$ 时,对应于标准的 Rosen-Zener 模型 [49]:

$$P = \frac{\sin^2\left(\frac{\pi T v_0}{2}\right)}{\cosh^2\left(\frac{\pi T \gamma_1}{2}\right)}. \quad (3.18)$$

当 $c \neq 0, \gamma_0 = 0$ 时,对应于非线性 Rosen-Zener 模型 [61,62]. 非线性会极大地影响跃迁动力学,导致许多有趣的现象. 对于弱非线性情况,在长扫描周期的参数范围会出现规则的矩形振荡,这种振荡的出现意味着在一个很宽的参数范围内,人们可以鲁棒地控制该系统,使其在两能级间进行完全的传递,且振荡周期随着非线性的加强而变大. 而对于强非线性情况,两能级之间的量子跃迁受非线性的影响更加严重,振荡模式不复存在,量子跃迁被完全阻塞.

3.2.2 无静态失谐情况

$\gamma_1 = 0$,意味着无静态失谐情况,Demkov-Kunike 跃迁简化为 Allen-Eberly 跃迁. 随着非线性的出现,系统的跃迁动力学发生了剧烈的变化,同时在这种情况下,

Schrödinger 方程不能解析求解 [71], 利用 4-5 阶 Runge-Kutta 法对方程进行数值求解. 在数值计算过程中, 选取最大耦合强度 v_0 作为标度, 因此, 对应的弱啁啾情况和强啁啾情况分别为 $\gamma_0/v_0 \ll 1$ 和 $\gamma_0/v_0 \gg 1$ [62].

首先讨论在弱啁啾情况下跃迁动力学的一些规律. 图 3.5 展示了不同粒子间相互作用下跃迁概率随扫描周期的变化情况. 最初所有的粒子分别被制备在 ψ_1 态 (上图) 和 ψ_2 态 (下图), 参数分别取 $v_0 = 1$, $\gamma_0 = 0.1$. 线性情况时 [图 3.5(a) 和 (a′)], 系统最初分别被制备在 ψ_1 态和 ψ_2 态上的跃迁动力学是完全对称的. 随着扫描周期 T 的增大, 跃迁概率的振荡逐渐消失, 在绝热情况下跃迁概率稳定到 1. 然而, 随着非线性的出现, 跃迁动力学发生了改变. 对于不同的初始态, 非线性导致了跃迁概率的不对称性. 对于弱非线性情况, 当系统初始被制备在 ψ_1 态上 [图 3.5 (b) 和 (c)], 随着扫描周期的增大, 非线性使跃迁概率的振荡逐渐减小直到消失.

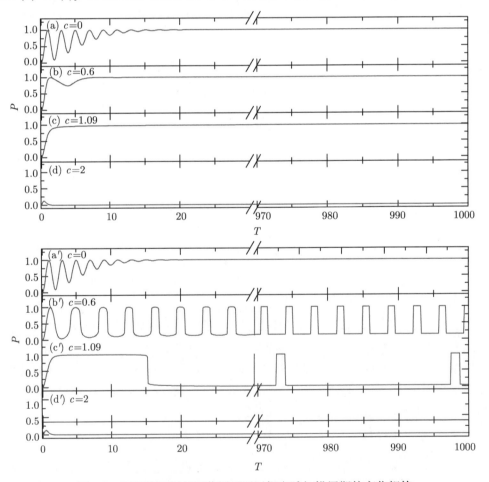

图 3.5 不同粒子间相互作用下跃迁概率随扫描周期的变化规律

3.2 非线性 Demkov-Kunike 跃迁

同时,随着非线性的增大,跃迁概率能很快稳定到值 1. 当系统初始制备在 ψ_2 上时 [图 3.5(b′) 和 (c′)],非线性使振荡更明显. 更为有趣的是,跃迁概率可以在从 0 到一个固定值之间振荡,在绝热极限下会出现一种矩形周期模式,其周期会随着非线性参数 c 的增大而增大,在整个参数范围跃迁概率在 $0 \sim 1$ 变化. 对于强相互作用 [图 3.5(d) 和 (d′)],非线性对两态之间的跃迁影响更大,不论初态在 ψ_1 还是 ψ_2 上,振荡模式完全消失,当扫描周期 $T > 2$ 时,量子跃迁被阻塞,跃迁概率变为 0. 在图 3.6 中也显示了不同扫描情况时跃迁概率随非线性相互作用 c 的变化情况,参数分别取 $v_0 = 1, \gamma_0 = 0.1$. 在快速扫描情况下 [图 3.6($T = 1$)],不论初始在哪个态上,非线性总是抑制了跃迁,这是由非线性引起的相互作用能和耦合强度引起的跳跃能量二者之间的相互竞争引起的. 当非线性相互作用 c 为主导时,相互作用能增加了,从而两态之间的跃迁被抑制[61].

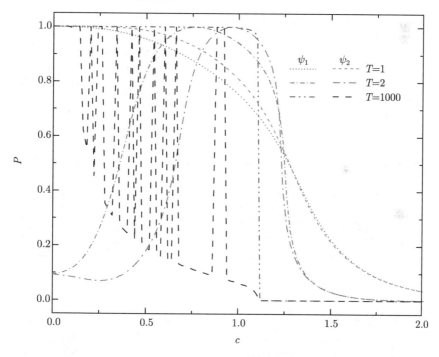

图 3.6 不同初始模态跃迁概率在不同扫描情况下随非线性相互作用的变化规律

进一步研究了非线性 Demkov-Kunike 跃迁过程的一般性质. 图 3.7 展示了在不同啁啾参数下 Demkov-Kunike 跃迁概率关于非线性参数和扫描周期的等高线图,图中最初所有的粒子分别被制备在 ψ_1(左栏) 和 ψ_2(右栏) 上. 图 3.7 揭示了 Demkov-Kunike 跃迁的主要特征: ① 跃迁概率存在三个完全不同的区域: 完全跃迁区、部分跃迁区和跃迁完全阻塞区. ② 不同的初始态下, 跃迁概率具有不对称性, 见图 3.7 左右两栏. 当系统初始放在 ψ_1 上时, 对于弱非线性相互作用, 随着扫描周期的增大, 跃迁概率能很快稳定在值 1, 同时在小的扫描周期时, 振荡会随着 γ_0 增加而消失 [图 3.7(a)、(c) 和 (e)]. 对大的扫描周期, 跃迁概率仅和非线性相互作用有关 [图 3.7(b)、(d) 和 (f)]. 然而, 如果系统初始制备在 ψ_2 上, 跃迁概率将在很大的参数范围内在 0 和 1 之间变化 [图 3.7(a′)、(c′) 和 (e′)]. ③ 对于强相互作用, 不论初始态在哪里, 跃迁概率都变为 0, 意味着跃迁被完全阻塞. 和弱啁啾参数不同, 对于强的啁啾参数, 只要相互作用足够强, 跃迁概率总能变为 0, 跃迁总会被阻塞.

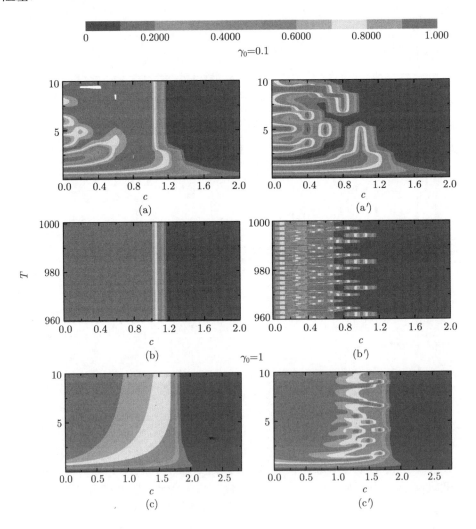

3.2 非线性 Demkov-Kunike 跃迁

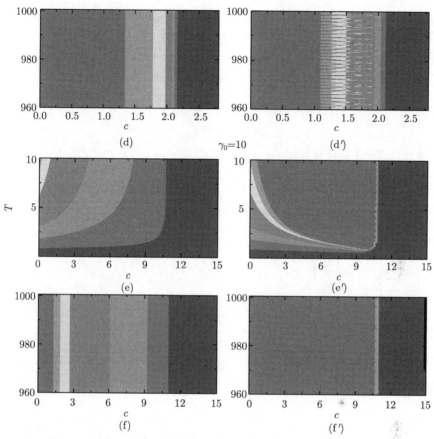

图 3.7 不同啁啾参数下 Demkov-Kunike 跃迁概率关于非线性参数和扫描周期的等高线图

为了进一步理解上述现象, 分析了系统的本征能级结构. 由非线性情况下的绝热理论, 绝热极限下量子跃迁的性质完全由系统的能级结构和对应的本征态 (对应于经典哈密顿系统中不动点的稳定性) 决定. 系统的本征态为

$$H(t)\begin{pmatrix} a \\ b \end{pmatrix} = \mu \begin{pmatrix} a \\ b \end{pmatrix}, \tag{3.19}$$

其中, μ 是系统的化学势 [37], 表示向系统中加入一个粒子所需的能量. 求解上述方程组, 同时利用粒子数守恒条件 $|a|^2 + |b|^2 = 1$, 会得到系统的化学势 μ 和本征态 (a, b). 相应的本征能可由 $\varepsilon = \mu - c/2(|a|^4 + |b|^4)$ 给出. 以弱啁啾情况 ($\gamma_0 = 0.1$) 为例, 如图 3.8 所示, 图中参数 v_0 和 γ_0 的取值与图 3.5 相同. 图中 (a)、(b) 与 (c)、(d) 分别代表无相互作用 (线性情况)、弱相互作用和强相互作用, 圆圈和三角分别表示初态 ψ_1 和 ψ_2. 其中, 虚线代表不稳定能级 [61]. 其稳定性可以通过计算方程 (3.19) 线性化后的 Hamiltonian-Jacobi 矩阵的本征值来加以说明. Hamiltonian-Jacobi 矩

阵的本征值可以是实数、复数或纯虚数. 只有纯虚数对应的本征态才是稳定的, 其他的都是不稳定的 [62].

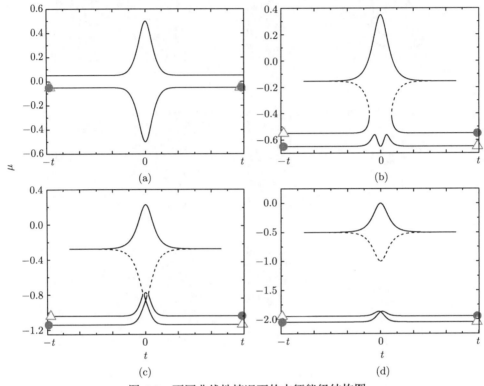

图 3.8 不同非线性情况下的本征能级结构图

(a) $c = 0$; (b) $c = 0.6$; (c) $c = 1.09$; (d) $c = 2$

由非线性导致的能级结构的变化显然会影响系统的量子跃迁. 在线性情况下 [图 3.8(a)], 初始时刻分别被制备在 ψ_1 态和 ψ_2 态的系统, 本征能级结构是对称的, 这导致了对称的跃迁动力学 [图 3.5(a) 和 (a′)]. 然而在弱相互作用下, 对于不同的初始态, 本征能级结构发生了变化 [图 3.8(b) 和 (c)]. 当初始系统被制备在 ψ_1 态时, 系统沿着本征能级绝热跟随, 直到能级结构被破坏, 并且在适当的非线性参数下出现了环状结构. 正是这个环状结构使得跃迁概率即使在绝热情况下也不为 1 [图 3.5(c) 和图 3.6($T = 1000$)]. 当初始系统被制备在 ψ_2 态时, 随着演化时间的增加, 由于在 $t = 0$ 附近能级突然中止, 前方已无路可走, 系统不再绝热跟随, 被迫跳跃到其他能级上 [图 3.8(b) 和 (c)], 从而使得跃迁概率出现振荡 [图 3.5(b′) 和 (c′)]. 在强相互作用下, 非线性使本征能级结构完全破坏, 但是初始时刻分别被制备在 ψ_1 态和 ψ_2 态系统的本征能级结构具有对称性 [图 3.8(d)], 因此其跃迁动力学相似 [图 3.5(d) 和 (d′)].

3.2 非线性 Demkov-Kunike 跃迁

在一些特殊情况下，如大扫描周期 T 和强相互作用 c 时，通过稳相近似[35] 也能获得一些解析结果. 大周期时，Allen-Eberly 模型简化为 Landau-Zener 模型，由文献 [35]，可以获得跃迁概率的一个封闭方程:

$$\frac{1}{P} = \frac{2\gamma_0}{\pi T v_0^2} + \frac{2c}{\pi v_0}\sqrt{P}. \tag{3.20}$$

绝热极限下，$P = \left(\frac{\pi v_0}{2c}\right)^{\frac{2}{3}}$. 该结果也可以直接通过解 Schrödinger 方程验证，二者的对比如图 3.9 所示，系统初始态为 ψ_1，其他参数取为 $\gamma_0 = 1, v_0 = 0.1$.

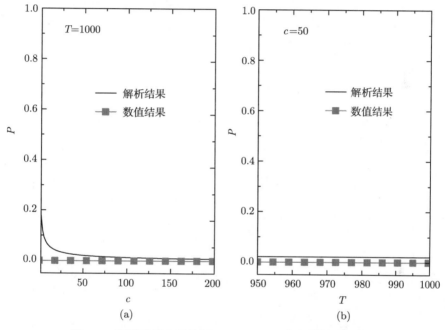

图 3.9 解析结果和直接解 Schrödinger 方程数值结果对比

3.2.3 存在静态失谐情况

本节研究静态失谐参数 ($\gamma_1 \neq 0$) 对跃迁动力学的影响. 图 3.10 展示了在非线性情况下不同静态失谐参数下 Demkov-Kunike 跃迁概率随扫描周期的变化规律. 图中参数分别取 $c = 0.6, v_0 = 1, \gamma_0 = 0.1$. 最初所有的粒子分别被制备在 ψ_1 态 (左栏) 和 ψ_2 态 (右栏). 随着静态失谐参数的出现，跃迁动力学发生了改变. 对于给定的非线性参数，无论初始时刻系统被制备在 ψ_1 态还是 ψ_2 态，跃迁概率都随着不同的静态失谐参数发生变化.

与无静态失谐情况比较，不同的静态失谐参数对不同初始态的跃迁动力学的影响是不同的. $\gamma_1 > 0$ 时，主要影响从 ψ_2 态开始演化的系统 [图 3.10 (d) 和 (f)].

随着静态失谐参数 γ_1 的增大, 跃迁概率的矩形振荡逐渐消失, 跃迁概率趋于零 [图 3.9(b)→ (d)→ (f)]. 当 $\gamma_1 < 0$ 时, 对于初始态 ψ_1 [图 3.10(g) 和 (i)], 随着静态失谐参数 γ_1 降低, 跃迁概率会趋于 0. 然而对于初始态 ψ_2, 随着静态失谐参数 γ_1 的降低, 即 $|\gamma_1|$ 增加, 跃迁概率的矩形振荡仍然存在而且其幅度会变小 [图 3.10(j)].

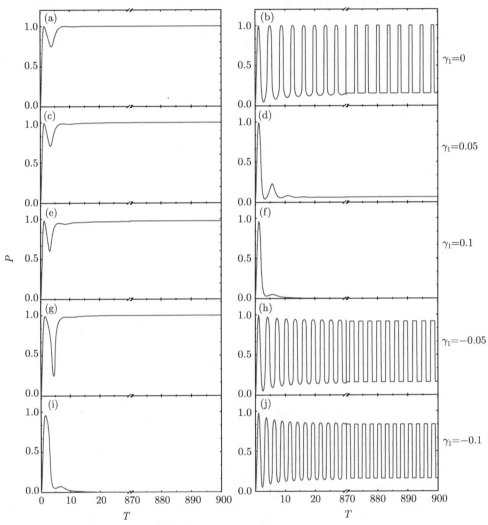

图 3.10 非线性情况下不同静态失谐参数下 Demkov-Kunike 跃迁概率随扫描周期的变化规律

与此同时, 发现当 $|\gamma_1| \geqslant \gamma_0$ 时, 正的 γ_1 初态在 ψ_2 与负的 γ_1 初态在 ψ_1 时系统的跃迁动力学相同. 这里假定所有的相互作用取正值, 对应系统中为排斥相互作用, 研究也表明, 吸引相互作用时从 ψ_1 到 ψ_2 的跃迁和排斥相互作用时从 ψ_2 到 ψ_1

的跃迁相同[61].

同样可以通过分析系统的本征能级结构理解上述结果. 图 3.11 展示了在非线性情况下 ($c = 0.6$), 不同的静态失谐参数对应的本征能级结构, 其中圆圈和三角分别表示 ψ_1 态和 ψ_2 态, 虚线表示不稳定能级. 图 3.11(b) 和 (c) 中, 部分能级结构是简并的, 直接导致了不同静态失谐参数下跃迁动力学的差异. 图 3.11 (b) 和 (c) 的能级结构具有反向对称性, 这也直接导致不同初始状态下跃迁动力学的性质.

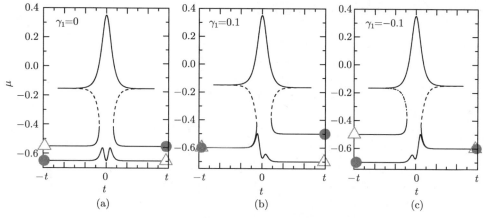

图 3.11 非线性情况下 ($c = 0.6$) 不同静态失谐参数对应的本征能级结构图

3.2.4 应用和讨论

Demkov-Kunike 模型能描述许多真实的物理系统, 如分子波包的激发过程[73]、量子绝热态的制备[74]、耦合波导管中布居数转移[57] 以及光晶格 Bloch 能带间的玻色-爱因斯坦凝聚等. 特别在光晶格玻色-爱因斯坦凝聚中, 其跃迁动力学可以很好地通过上述方程描述[18]. 该系统中, γ 代表能级差, v 代表耦合强度, 它们可以通过外场调节. 非线性参数 $c = \pi n_0 a_s / k_L^2$ 同样可以通过 Feshbach 共振适当地调节, 这里 n_0 是凝聚态原子的平均密度, a_s 是原子间 s 波的散射长度, k_L 是外场的波数. 在一个钠原子典型的实验中, 平均密度可取为 $n_0 = 3 \times 10^{21}$ m^{-3}, 当散射长度 $a_s = a_0 = 2.75$nm (a_0 是非共振下的扫描长度) 时, 这个非线性参数 $c = 0.268$[18,75], 若散射长度是 $a_s = 5a_0$ [76], 则非线性参数 $c = 1.34$. 当然也可以通过改变凝聚态原子的平均密度 n_0 和外场的波数 k_L 来获得非线性参数 c, 而脉冲宽度 T 也可以通过两个外部场调节[71,77].

另外, 有望用于观察非线性 Demkov-Kunike 跃迁行为的物理系统是双势阱中的玻色-爱因斯坦凝聚[28]. 在这个系统中, 波函数可以写出两个近对立阱中基态的线性叠加, 光学双势阱可以通过在磁阱的中心叠加一束蓝失谐的激光来实现. 在这种情况中, γ 代表两个阱零点能之差, c 表示原子间相互作用, 可以通过 Feshbach 任意调节，v 表示两阱势垒的高度，可以通过调节蓝失谐激光的强度有效控制[62].

总之，本节研究了存在粒子间相互作用的非线性两能级系统中 Demkov-Kunike 跃迁动力学，主要讨论了粒子间相互作用对跃迁概率的影响. 不同于线性量子系统，粒子间相互作用的存在使 Demkov-Kunike 模型的量子系统表现出许多有趣的现象. 特别是，非线性导致的跃迁概率在不同初始状态的不对称性. 这些非线性系统中 Demkov-Kunike 跃迁的规律，不仅有助于理解非线性跃迁动力学，同时对实现量子系统的外场控制等都具有重要的意义[71,77].

参 考 文 献

[1] LANDAU L D. On the theory of transfer of energy at collisions II[J]. Phys. Z. Sowjetunion, 1932, 2: 46-51.

[2] ZENER C. Non-adiabatic crossing of energy levels[J]. Proc. R. Soc. London, 1932, 137: 696-702.

[3] STÜCKELBERG E C G. Theory of inelastic collisions between atoms (Theory of inelastic collisions between atoms, using two simultaneous differential equations)[J]. Helv. Phys. Acta, 1932, 5: 369-422.

[4] MAJORANA E. Atomi orientati in campo magnetico variabile[J]. Nuovo Cimento, 1932, 9: 43-50.

[5] WERNSDORFER W, SESSOLI R. Quantum phase interference and parity effects in magnetic molecular clusters[J]. Science, 1999, 284(5411): 133-135.

[6] FÖLDI P, BENEDICT M G, PEREIRA J M, et al. Dynamics of molecular nanomagnets in time-dependent external magnetic fields: Beyond the Landau-Zener-Stückelberg model[J]. Phys. Rev. B, 2007, 75: 104430.

[7] CHILD S. Molecular Collision Theory[M]. New York: Dover, 1974.

[8] SPREEUW R J C, VAN DRUTEN N J, BEIJERSBERGEN M W, et al. Classical realization of a strongly driven two-level system[J]. Phys. Rev. Lett., 1990, 65: 2642-2645.

[9] RUBBMAR J R, KASH M M, LITTMAN M G, et al. Dynamical effects at avoided level crossings: A study of the Landau-Zener effect using Rydberg atoms[J]. Phys. Rev. A, 1981, 23: 3107-3117.

[10] JONA-LASINIO M, MORSCH O, CRISTIANI M, et al. Asymmetric Landau-Zener tunneling in a periodic potential[J]. Phys. Rev. Lett., 2003, 91: 230406.

[11] WOO S J, CHOI S, BIGELOW N P. Controlling quasiparticle excitations in a trapped Bose-Einstein condensate[J]. Phys. Rev. A, 2005, 72: 021605.

[12] LIU J, LIU B, FU L B. Many-body effects on nonadiabatic Feshbach conversion in bosonic systems[J]. Phys. Rev. A, 2008, 78(1): 013618.

[13] LIU J, FU L B, LIU B, et al. Role of particle interactions in the Feshbach conversion of fermionic atoms to bosonic molecules[J]. New J. Phys., 2008, 10(12): 123018.

[14] SIBILLE A, PALMIER J F, LARUELLE F. Zener interminiband resonant breakdown in superlattices[J]. Phys. Rev. Lett., 1998, 80: 4506-4509.

[15] MULLEN K, BEN-JACOB E, SCHUSS Z. Combined effect of Zener and quasiparticle transitions on the dynamics of mesoscopic Josephson junctions[J]. Phys. Rev. Lett., 1988, 60: 1097-1100.

[16] KHOMERIKI R, RUFFO S. Nonadiabatic Landau-Zener tunneling in waveguide arrays with a step in the refractive index[J]. Phys. Rev. Lett., 2005, 94: 113904.

[17] HIOE F, CARROLL C. Analytic solutions to the two-state problem for chirped pulses[J]. J. Opt. Soc. Am. B, 1985, 2: 497-502.

[18] WU B, NIU Q. Nonlinear Landau-Zener tunneling[J]. Phys. Rev. A, 2000, 61: 023402.

[19] LIU J, FU L, OU B Y, et al. Theory of nonlinear Landau-Zener tunneling[J]. Phys. Rev. A, 2002, 66: 023404.

[20] ZOBAY O, GARRAWAY B M. Time-dependent tunneling of Bose-Einstein condensates[J]. Phys. Rev. A, 2000, 61: 033603.

[21] VITANOV N V, GARRAWAY B M. Landau-Zener model: Effects of finite coupling duration[J]. Phys. Rev. A, 1996, 53: 4288-4304.

[22] POKROVSKY V L, SINITSYN N A. Landau-Zener transitions in a linear chain[J]. Phys. Rev. B, 2002, 65: 153105.

[23] OSTROVSKY V N, VOLKOV M V, HANSEN J P, et al. Four-state nonstationary models in multistate Landau-Zener theory[J]. Phys. Rev. B, 2007, 75: 014441.

[24] SHIMSHONI E, STERN A. Dephasing of interference in Landau-Zener transitions[J]. Phys. Rev. B, 1993, 47: 9523-9536.

[25] CHEN Y A, HUBER S D, TROTZKY S, et al. Many-body Landau-Zener dynamics in coupled one-dimensional Bose liquids[J]. Nature Phys., 2010, 7(1): 61-67.

[26] LIU J, WU B, FU L, et al. Quantum step heights in hysteresis loops of molecular magnets[J]. Phys. Rev. B, 2002, 65(22): 224401.

[27] SMERZI A, FANTONI S, GIOVANAZZI S, et al. Quantum coherent atomic tunneling between two trapped Bose-Einstein condensates[J]. Phys. Rev. Lett., 1997, 79: 4950-4953.

[28] ALBIEZ M, GATI R, FÖLLING J, et al. Direct observation of tunneling and nonlinear self-trapping in a single bosonic Josephson junction[J]. Phys. Rev. Lett., 2005, 95: 010402.

[29] KHOMERIKI R. Nonlinear Landau-Zener tunneling in coupled waveguide arrays[J]. Phys. Rev. A, 2010, 82: 013839.

[30] WITTHAUT D, GRAEFE E M, KORSCH H J. Towards a generalized Landau-Zener formula for an interacting Bose-Einstein condensate in a two-level system[J]. Phys. Rev. A, 2006, 73: 063609.

[31] GARANIN D A. Landau-Zener-Stueckelberg effect in a model of interacting tunneling systems[J]. Phys. Rev. B, 2003, 68: 014414.

[32] VITANOV N V, SUOMINEN K A. Nonlinear level-crossing models[J]. Phys. Rev. A, 1999, 59: 4580-4588.

[33] GARRAWAY B M, VITANOV N V. Population dynamics and phase effects in periodic level crossings[J]. Phys. Rev. A, 1997, 55: 4418-4432.

[34] GARANIN D A, SCHILLING R. Effects of nonlinear sweep in the Landau-Zener-Stueckelberg effect[J]. Phys. Rev. B, 2002, 66: 174438.

[35] DOU F Q, LI S C, CAO H. Combined effects of particle interaction and nonlinear sweep on Landau-Zener transition[J]. Phys. Lett. A, 2011, 376(1): 51-55.

[36] LIU W V, WILCZEK F. Interior gap superfluidity[J]. Phys. Rev. Lett., 2003, 90: 047002.

[37] WU B, NIU Q. Superfluidity of Bose-Einstein condensate in an optical lattice: Landau-Zener tunnelling and dynamical instability[J]. New J. Phys., 2003, 5(1): 104.

[38] JONA-LASINIO M, MORSCH O, CRISTIANI M, et al. Nonlinear effects in periodic potentials: asymmetric Landau-Zener tunnelling of a Bose-Einstein condensate[J]. Las. Phys. Lett., 2004, 1(3): 147-153.

[39] LIU J, FU L, OU B Y, et al. Theory of nonlinear Landau-Zener tunneling[J]. Phys. Rev. A, 2002, 66(2): 023404.

[40] REICHL L E. The Transition to Chaos: Conservative Classical Systems and Quantum Manifestations[M]. New York: Springer, 2004.

[41] DALFOVO F, GIORGINI S, PITAEVSKII L P, et al. Theory of Bose-Einstein condensation in trapped gases[J]. Rev. Mod. Phys., 1999, 71: 463-512.

[42] MCMAHON R J. Chemical reactions involving quantum tunneling[J]. Science, 2003, 299(5608): 833-834.

[43] BALANTEKIN A, TAKIGAWA N. Quantum tunneling in nuclear fusion[J]. Rev. Mod. Phys., 1998, 70(1): 77.

[44] ROSEN N, ZENER C. Double Stern-Gerlach experiment and related collision phenomena[J]. Phys. Rev., 1932, 40(4): 502-507.

[45] ALLEN L, EBERLY J H. Optical Resonance and Two-Level Atoms[M]. New York: Dover, 1987.

[46] HIOE F. Solution of Bloch equations involving amplitude and frequency modulations[J]. Phys. Rev. A, 1984, 30(4): 2100.

[47] BAMBINI A, BERMAN P R. Analytic solutions to the two-state problem for a class of coupling potentials[J]. Phys. Rev. A, 1981, 23: 2496-2501.

[48] SIMEONOV L S, VITANOV N V. Exactly solvable two-state quantum model for a pulse of hyperbolic-tangent shape[J]. Phys. Rev. A, 2014, 89(4): 043411.

[49] LACOUR X, GUERIN S, YATSENKO L, et al. Uniform analytic description of dephasing effects in two-state transitions[J]. Phys. Rev. A, 2007, 75(3): 033417.

[50] DEMKOV Y N, KUNIKE M. Hypergeometric models for the two-state approximation in collision theory[J]. Vestn. Leningr. Univ., 1969,16(3): 39.

[51] SUOMINEN K A, GARRAWAY B. Population transfer in a level-crossing model with two time scales[J]. Phys. Rev. A, 1992, 45(1): 374.

[52] OTA Y, KONDO Y. Composite pulses in NMR as nonadiabatic geometric quantum gates[J].

Phys. Rev. A, 2009, 80(2): 024302.

[53] RUSCHHAUPT A, MUGA J. Adiabatic interpretation of a two-level atom diode, a laser device for unidirectional transmission of ground-state atoms[J]. Phys. Rev. A, 2006, 73(1): 013608.

[54] WEITZ M, YOUNG B C, CHU S. Atomic interferometer based on adiabatic population transfer[J]. Phys. Rev. Lett., 1994, 73(19): 2563.

[55] BERGMANN K, THEUER H, SHORE B W. Coherent population transfer among quantum states of atoms and molecules[J]. Rev. Mod. Phys., 1998, 70: 1003-1025.

[56] TOROSOV B, VITANOV N. Coherent control of a quantum transition by a phase jump[J]. Phys. Rev. A, 2007, 76(5): 053404.

[57] PAUL K, SARMA A K. Shortcut to adiabatic passage in a waveguide coupler with a complex-hyperbolic-secant scheme[J]. Phys. Rev. A, 2015, 91(5): 053406.

[58] WU B, LIU J. Commutability between the semiclassical and adiabatic limits[J]. Phys. Rev. Lett., 2006, 96: 020405.

[59] ISHKHANYAN A. Generalized formula for the Landau-Zener transition in interacting Bose-Einstein condensates[J]. Europhys. Lett., 2010, 90(3): 30007.

[60] TRIMBORN F, WITTHAUT D, KEGEL V, et al. Nonlinear Landau-Zener tunneling in quantum phase space[J]. N. J. Phys., 2010, 12(5): 053010.

[61] YE D F, FU L B, LIU J. Rosen-Zener transition in a nonlinear two-level system[J]. Phys. Rev. A, 2008, 77(1): 013402.

[62] 叶地发. 玻色爱因斯坦凝聚体的量子相干调控 [D]. 绵阳: 中国工程物理研究院, 2008.

[63] LI S C, FU L B, DUAN W S, et al. Nonlinear Ramsey interferometry with Rosen-Zener pulses on a two-component Bose-Einstein condensate[J]. Phys. Rev. A, 2008, 78(6): 063621.

[64] FU L B, YE D F, LEE C, et al. Adiabatic Rosen-Zener interferometry with ultracold atoms[J]. Phys. Rev. A, 2009, 80(1): 013619.

[65] JONA-LASINIO M, MORSCH O, CRISTIANI M, et al. Asymmetric Landau-Zener tunneling in a periodic potential[J]. Phys. Rev. Lett., 2003, 91: 230406.

[66] JONA-LASINIO M, MORSCH O, CRISTIANI M, et al. Nonlinear effects in periodic potentials: asymmetric Landau-Zener tunnelling of a Bose-Einstein condensate[J]. Laser. Phys. Lett., 2004, 1(3): 147.

[67] MORALES-MOLINA L, AREVALO E. Accurate control of a Bose-Einstein condensate by managing the atomic interaction[J]. Phys. Rev. A, 2010, 82(1): 013642.

[68] ISHKHANYAN A, JOULAKIAN B, SUOMINEN K A. Two strong nonlinearity regimes in cold molecule formation[J]. Eur. Phys. J. D-Atomic, Molecular, Optical and Plasma Physics, 2008, 48(3): 397-404.

[69] SOKHOYAN R, AZIZBEKYAN H, LEROY C, et al. Demkov-Kunike model for cold atom association: Weak interaction regime[J]. Journal of Contemporary Physics (Armenian Academy of Sciences), 2009, 44(6): 272-277.

[70] SOKHOYAN R. Demkov-Kunike model in cold molecule formation: the fast resonance sweep regime of the strong interaction limit[J]. Journal of Contemporary Physics (Armenian Academy

of Sciences), 2010, 45(2): 51-57.

[71] DOU F Q, CAO H, LIU J, et al. High-fidelity composite adiabatic passage in nonlinear two-level systems[J]. Phys. Rev. A, 2016, 93(4): 043419.

[72] FENG P, WANG W Y, SUN J A, et al. Demkov-Kunike transition dynamics in a nonlinear two-level system[J]. Nonlinear Dyn., 2018, 91: 2477-2484.

[73] PALOVIITA A, SUOMINEN K A, STENHOLM S. Molecular excitation by chirped pulses[J]. J. Phys. B: Atomic, Molecular and Optical Physics, 1995, 28(8): 1463.

[74] LARSON J, STENHOLM S. Adiabatic state preparation in a cavity[J]. J. Mod. Opt., 2003, 50(11): 1663-1678.

[75] STAMPER-KURN D, ANDREWS M, CHIKKATUR A, et al. Optical confinement of a Bose-Einstein condensate[J]. Phys. Rev. Lett., 1998, 80(10): 2027.

[76] STENGER J, INOUYE S, CHIKKATUR A P, et al. Bragg spectroscopy of a Bose-Einstein condensate[J]. Phys. Rev. Lett., 1999, 82: 4569-4573.

[77] GUÉRIN S, HAKOBYAN V, JAUSLIN H. Optimal adiabatic passage by shaped pulses: Efficiency and robustness[J]. Phys. Rev. A, 2011, 84(1): 013423.

第 4 章 高保真度量子操控

第 3 章研究了广义非线性 Landau-Zener 隧穿和非线性 Demkov-Kunike 跃迁动力学. 本章从另一个不同的层面研究隧穿行为, 考虑系统如何从一个初态保持高保真度到达终态, 这在量子信息与量子计算中是非常有用的, 涉及超绝热量子驱动、超快量子控制以及复合绝热通道技术的布居数转移三个方面.

4.1 高保真度超绝热量子驱动

量子绝热过程是量子态控制中的一个重要的工具. 量子态控制的目标是以一种优化的方式操控到所需要的量子态[1-3]. 绝热演化过程中, 系统会跟随哈密顿量的一个本征态, 即如果初始时刻, 系统制备在一个本征态上, 之后系统将沿着该本征态演化. 最近几十年, 基于绝热动力学, 许多不同领域, 如化学反应的控制、激光冷却、核磁共振、量子信息以及量子光学中快的布居数转移等被广泛研究并在实验上得以实现[4-7]. 然而, 通常这样一个绝热过程因为要满足绝热条件, 也许太慢了, 并且几乎所有的绝热技术, 其布居数都不能实现完全转移, 即保真度接近或者小于 1[8], 所以人们提出了许多方法去加速绝热过程, 并提高量子操控过程的保真度[9-15].

在所有流行的方法中, 超绝热 (superadiabatic, 也称 transitionless 或 counter-diabatic) 量子驱动[11-13,16,17] 和绝热捷径 (shortcut to adiabaticity) 技术[14,18] 是加速绝热量子行为的两个重要工具. 前者通过构造一个附加场 (哈密顿量) 去抑制本征能级之间的非绝热跃迁, 从而确保极好的绝热跟随; 后者运用 Lewis-Riesenfeld (LR) 不变量来加速量子过程. 尽管它们形式不同, 但是已经证明通过调节参考哈密顿量, 二者是完全等价的[14]. 最近, 这些方法已经扩展到许多量子系统[19-29]. 对于超绝热技术, 实验上已经在加载在加速光晶格的 BEC 系统[30,31]、钻石单 Nitrogen-vacancy 色心的电子自旋系统[32]、大单光子失谐的冷原子系统[33]、俘获离子的连续变量等系统[34] 以及固态 Lamda 系统[35] 中被实现.

两能级系统是一个基本模型, 在量子力学中起着重要作用. 尽管在自然界中纯粹的两能级系统不太多, 然而它能作为许多物理领域的基本模型, 用以描述众多物理现象. 同时, 许多涉及多能级态和复杂模式的模型通常也能用一个简化的等效两能级模型去理解. 两能级系统中优化控制问题已经有很长的研究历史, 如著名的 Landau-Zener、Rosen-Zener、Demkov-Kunike、Roland-Cerf 以及复合脉冲等模型已经被研究了很多年[36–42]. 在绝热量子操控框架下, 高保真度超绝热量子驱动也已经在广义 Landau-Zener 模型、Allen-Eberly 模型以及 Tangent 模型等中被研究[30,31].

本节研究啁啾高斯两能级模型和 Demkov-Kunike 模型的超绝热量子驱动问题[43,44]. 通过适当选择附加场, 绝热操控效率大大提高. 附加场能完全抵消非绝热耦合, 加速绝热动力学过程. 几乎在所有的演化时刻, 都能实现系统的绝热跟随, 系统保持在绝热基态上. 4.1.1 小节简要介绍两能级系统中的超绝热量子驱动技术. 4.1.2 小节讨论超绝热技术在啁啾高斯模型中的应用. 4.1.3 小节讨论超绝热技术在 Demkov-Kunike 模型中的应用, 通过改变附加场强度, 研究附加场大小对超绝热过程的影响.

4.1.1 模型和超绝热技术

外场控制的两能级系统由下述无量纲化的 Schrödinger 方程来描述:

$$i\frac{\partial}{\partial t}\begin{pmatrix} a \\ b \end{pmatrix} = H(t) \begin{pmatrix} a \\ b \end{pmatrix}, \tag{4.1}$$

其中, 哈密顿量为

$$H(t) = \gamma(t)\hat{\sigma}_z + v(t)\hat{\sigma}_x, \tag{4.2}$$

其中, a 和 b 分别是非绝热态 $|0\rangle$ 和 $|1\rangle$ 的概率幅, 同样总概率 $|a|^2 + |b|^2$ 守恒, 并且设为 1; $\hat{\sigma}_x$ 和 $\hat{\sigma}_z$ 是泡利矩阵; $\gamma(t)$ 和 $v(t)$ 分别是两个非绝热能级之间的能级差和耦合强度.

上述系统有绝热本征态 $|\psi_\pm(t)\rangle$,

$$H(t)|\psi_\pm(t)\rangle = \varepsilon_\pm(t)|\psi_\pm(t)\rangle, \tag{4.3}$$

其中, 本征值 $\varepsilon_\pm(t) = \pm\sqrt{\gamma^2 + v^2}$, 下标 $-$ 和 $+$ 分别代表基态和激发态. 本征值的差 $\varepsilon(t) = \varepsilon_+(t) - \varepsilon_-(t) = 2\sqrt{\gamma^2 + v^2}$, 定义为能级分裂. 通过一个幺正变换, 即绝热基下, 系统哈密顿能够对角化. 该绝热基记为 (A, B), 则有

$$\begin{pmatrix} A \\ B \end{pmatrix} = U_0^{-1}(t) \begin{pmatrix} a \\ b \end{pmatrix}, \tag{4.4}$$

其中, U_0 是旋转矩阵, 形式为

$$U_0 = \begin{pmatrix} -\sin\theta & \cos\theta \\ \cos\theta & \sin\theta \end{pmatrix}. \tag{4.5}$$

其中, 混合角 $\theta = \frac{1}{2}\arctan[v(t)/\gamma(t)]$. 绝热基下, 系统哈密顿变为

$$H'(t) = U_0^{-1}H(t)U_0 - \mathrm{i}U_0^{-1}\dot{U}_0, \tag{4.6}$$

其中, "·" 表示对时间 t 的导数. 第一项是对角部分, 第二项是非对角部分, 看作非绝热修正. 绝热基下, Schrödinger 方程写为

$$\mathrm{i}\frac{\partial}{\partial t}\begin{pmatrix} A \\ B \end{pmatrix} = H'(t)\begin{pmatrix} A \\ B \end{pmatrix} = \begin{pmatrix} \varepsilon_- & -\mathrm{i}\dot{\theta} \\ \mathrm{i}\dot{\theta} & \varepsilon_+ \end{pmatrix}\begin{pmatrix} A \\ B \end{pmatrix}. \tag{4.7}$$

当哈密顿量中的非绝热耦合相比于能级分裂可以忽略时, 系统会发生绝热演化. 数学上, 绝热演化要求哈密顿量 (4.7) 中的副对角元相比于对角元可以忽略, 即 $|\dot{\theta}| \ll \varepsilon$, 这就是所谓的绝热条件[11]. 布居数的转移效率由于绝热条件大大被限制了, 而且要求演化过程非常慢. 当绝热条件不能满足时, 由于哈密顿量中非绝热项的存在, 布居数完全转移不会发生. 为了克服这些约束, 人们在系统中构造了一个附加哈密顿量 H_{cd} (也称 counter-diabatic 场) 去抵消单独沿 H 演化时的非绝热部分[11-13]. 这样确保系统沿着 $H+H_{\mathrm{cd}}$ 演化过程中能无跃迁地绝热跟随, 甚至在有限的持续时间内都保持在 H 的绝热基态上, 跃迁概率为 1. 一般而言, $H_{\mathrm{cd}} = \mathrm{i}\dot{U}_0U_0^{-1}$. 对于两能级系统 (4.1), H_{cd} 的形式为[11-13]

$$H_{\mathrm{cd}}(t) = \frac{\partial \theta}{\partial t}\hat{\sigma}_y, \tag{4.8}$$

其中, $\hat{\sigma}_y$ 是泡利矩阵.

为了研究布居数转移效率的稳定性, 改变附加场大小, 总哈密顿量形式如下:

$$\begin{aligned}
H_{\mathrm{tot}}(t) &= H + (1+\lambda)H_{\mathrm{cd}} \\
&= \gamma(t)\hat{\sigma}_z + v(t)\hat{\sigma}_x + (1+\lambda)\dot{\theta}\hat{\sigma}_y \\
&= \begin{pmatrix} \gamma(t) & v(t) - \mathrm{i}(1+\lambda)\dot{\theta} \\ v(t) + \mathrm{i}(1+\lambda)\dot{\theta} & -\gamma(t) \end{pmatrix},
\end{aligned} \tag{4.9}$$

其中, λ 表示附加场比率[45]. 当 $\lambda = -1$ 时, 对应于最初没有附加场时原始的哈密顿量 H, 而当 $\lambda = 0$ 时, 意味着在原哈密顿量基础上增加了一个附加场. 很明显, λ 的大小决定着附加场强度大小. 也能将式 (4.9) 写成一个有特定相位的等效耦合形式:

$$H_{\mathrm{tot}}(t) = \begin{pmatrix} \gamma(t) & v_{\mathrm{eff}}(t)\exp(-\mathrm{i}\phi) \\ v_{\mathrm{eff}}(t)\exp(\mathrm{i}\phi) & -\gamma(t) \end{pmatrix}, \tag{4.10}$$

其中, $v_{\mathrm{eff}}(t) = \sqrt{v^2(t) + [(1+\lambda)\dot{\theta}]^2}$. 为了抵消相位的依赖性, 做下述变换[22]:

$$U_1 = \begin{pmatrix} \exp(-\mathrm{i}\phi/2) & 0 \\ 0 & \exp(\mathrm{i}\phi/2) \end{pmatrix}, \tag{4.11}$$

这样实际上提供了一组新的基, 这时哈密顿量变为

$$H_{\text{tot}}(t) = \begin{pmatrix} \gamma_{\text{eff}}(t) & v_{\text{eff}}(t) \\ v_{\text{eff}}(t) & -\gamma_{\text{eff}}(t) \end{pmatrix}, \tag{4.12}$$

其中, $\gamma_{\text{eff}}(t) = \gamma(t) - \dot{\phi}/2$, $\phi = \arctan[(1+\lambda)\dot{\theta}/v(t)]$. 这说明控制外场也能通过适当的变换 $\gamma \to \gamma_{\text{eff}}$ 和 $v \to v_{\text{eff}}$ 变为等效外场形式. 通过上述方法, 已实现了 Landau-Zener、Allen-Eberly 等模型中的绝热跟随的完全布居数转移 [30,31].

4.1.2 啁啾 Gaussian 模型中的高保真度超绝热量子驱动

具有 Gaussian 型时间包络和线性失谐的啁啾高斯模型是两能级系统中的一个典型模型, 目前已广泛应用于原子分子和光物理以及等离子体物理中 [46-51]. 实验上, 该模型也已经应用于调节带宽频率 (bandwidth frequency)、模拟大气中激光触发闪电过程、测量孤子分子的相结构以及实现强激光场超快相干控制等 [46,49,50]. 通过运用 Dykhne-Davis-Pechukas 方法, 人们已经获得该模型关于跃迁概率的解析近似解 [51], 同时研究了其布居数转移问题 [50,52,53]. 这些研究也表明, 只有在绝热条件完全满足时, 才能实现其绝热跟随. 然而, 这样的绝热跟随过程为了满足绝热条件需要一个非常漫长的过程, 并且即使实现了绝热跟随, 其保真度也不高 [51,52]. 本节将超绝热量子驱动技术应用于啁啾 Gaussian 模型, 其中耦合随时间变化是 Gaussian 形式, 而能级差是时间的线性函数 [51]:

$$v(t) = v_0 \exp\left[-\left(\frac{t}{T}\right)^2\right], \quad \gamma(t) = \alpha t, \tag{4.13}$$

其中, v_0 是耦合强度 (或者 Rabi 频率的峰值); T 是脉冲持续时间; α 代表扫描率 (或者啁啾率). 假定初始时刻 $t = t_{\text{ini}}$ 时, 系统制备在绝热基态 $|\psi_-(t_{\text{ini}})\rangle$ 上, 经过演化时间 $t_{\text{fin}} - t_{\text{ini}}$ 后, $t = t_{\text{fin}}$ 时, 终态为 $|\psi_{\text{fin}}\rangle$. 量子驱动的目标是实现一个绝热技术, 确保在所有时刻系统都能绝热跟随在瞬时绝热基态 $|\psi_-(t)\rangle$ 上. 该技术能以一种加速的方式驱动系统从初态 $|\psi_-(t_{\text{ini}})\rangle$ 到终态 $|\psi_{\text{fin}}\rangle$, 并且具有高保真度, 即终态 $|\psi_{\text{fin}}\rangle$ 尽可能在绝热基态 $|\psi_-(t_{\text{fin}})\rangle$ 上, 从而使保真度为 1. 这里保真度 F_{fin} 定义为

$$F_{\text{fin}} = |\langle \psi_{\text{fin}} | \psi_-(t_{\text{fin}}) \rangle|^2. \tag{4.14}$$

对于啁啾 Gaussian 模型，$\dot{\theta} = [\dot{v}(t)\gamma(t) - v(t)\dot{\gamma}(t)]/2[v^2(t) + \gamma^2(t)] = -\alpha v(t)(t^2/T^2 + 0.5)/(v^2(t) + \gamma^2(t))$. 等效能级差和等效耦合在不同附加场比率情况下的时间依赖关系如图 4.1 所示，图中其他参数值：$v_0 = 2, \alpha = 0.5, T = 1$. 不同附加场比率时终态保真度随耦合强度 v_0 的变化如图 4.2 所示. 为简单起见，所有的变量都是无量纲化的，用 T 去重新尺度化. 计算中取 $T = 1$，则 v_0 和 α 的单位分别是 $1/T$ 和 $1/T^2$. 所有的数值模拟中，时间为 $-20 \sim 20$.

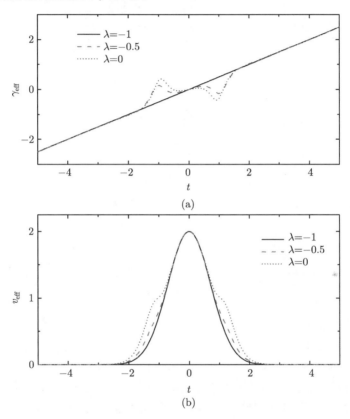

图 4.1　不同附加场比率时等效能级差和等效耦合的时间演化

(a) $\gamma_{\text{eff}}(t)$-t; (b) $v_{\text{eff}}(t)$-t

当 $\lambda = -1$ 时，对应于没有附加场的啁啾 Gaussian 模型，对于小的 v_0，终态保真度 F_{fin} 随着 v_0 单调增加，当 v_0 逐渐增加时，Rabi 类型的振荡出现 (除了非常大的 α 外). 随着附加场的出现，非绝热损失越来越小，振荡幅度也逐渐降低. 当 $\lambda = -0.5$ 时，终态保真度逐渐提高，直到 $\lambda = 0$，非绝热振荡完全消失，保真度保持在 1.

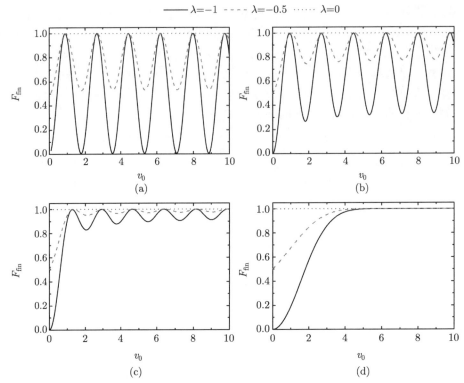

图 4.2 不同附加场比率和扫描率下终态保真度随耦合强度 v_0 的变化

(a) $\alpha = 0.05$; (b) $\alpha = 0.5$; (c) $\alpha = 1.5$; (d) $\alpha = 15$

图 4.3 中也说明了在不同附加场比率 λ 和耦合强度 v_0 时终态保真度随扫描率 α 的变化情况. 保真度随 α 没有出现振荡, 在 $\lambda = -1$ 时分成了几个不同的区域. 对于小的 α 值, F_{fin} 依赖于 v_0; 对于中间的 α 值, 保真度接近于 1; 对于非常大的 α 值, 保真度是逐渐降低的. 当 $\lambda = -0.5$ 时, 保真度逐渐变大; 当 $\lambda = 0$ 时, 意味着附加场完全打开, 在所有参数范围实现了超高保真度. 终态保真度随附加场比率 λ 的变化情况如图 4.4 所示. 可以发现, 附加场极大地提高了跃迁概率, 实现了高保真度, 并且保真度关于 λ 具有对称性, 在 $\lambda = 0$ 时, 保真度的值最大. 这是由于具有总哈密顿量 (4.9) 的 Schrödinger 在绝热基下为

$$\mathrm{i}\frac{\partial}{\partial t}\begin{pmatrix} A \\ B \end{pmatrix} = H'(t)\begin{pmatrix} A \\ B \end{pmatrix} = \begin{pmatrix} \varepsilon_- & -\mathrm{i}\lambda\dot{\theta} \\ \mathrm{i}\lambda\dot{\theta} & \varepsilon_+ \end{pmatrix}\begin{pmatrix} A \\ B \end{pmatrix}. \tag{4.15}$$

显然, 由方程 (4.15) 可以看出, 系统在 $\lambda = 0$ 时能完全地绝热跟随, 其跃迁概率关于 $\pm\lambda$ 完全对称.

4.1 高保真度超绝热量子驱动

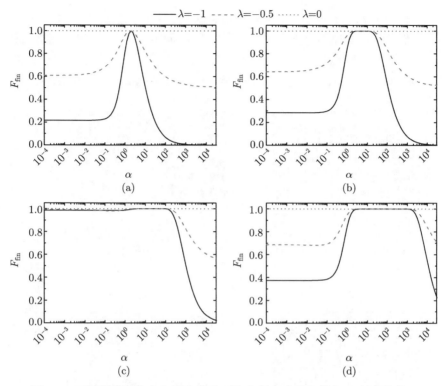

图 4.3 不同附加场比率和耦合强度下终态保真度随扫描率 α 的变化情况

(a) $v_0 = 1.5$; (b) $v_0 = 5$; (c) $v_0 = 15$; (d) $v_0 = 50$

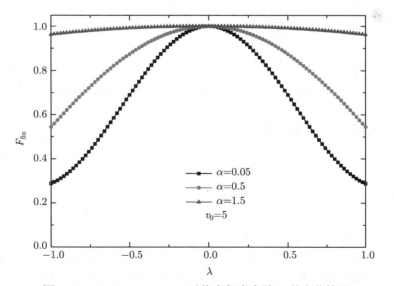

图 4.4 $\alpha = 0.05$、0.5、1.5 时终态保真度随 λ 的变化情况

为进一步研究该方法的有效性和参数鲁棒性,计算了在四个不同的 λ 值时终态保真度随扫描率 α 和耦合强度 v_0 的等密度图,如图 4.5 所示. 可以看到, 超绝热技术大大提高了关于参数 α 和 v_0 的鲁棒性, 并且实现了超高保真度 (误差低于量子计算的阈值 10^{-4}), 甚至在原先不满足绝热条件时的参数区域也如此.

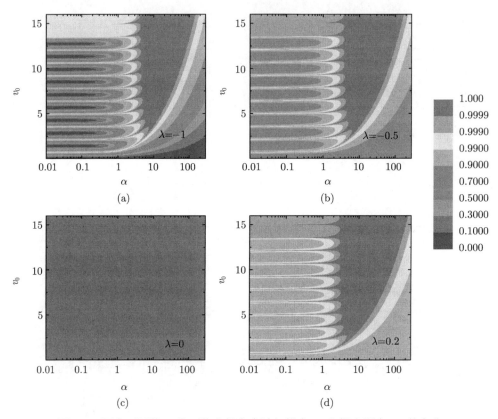

图 4.5 四个不同的 λ 值下终态保真度随扫描率 α 和耦合强度 v_0 的变化

为了刻画该方法的有效性, 定义了一个系统的瞬时保真度:

$$F(t) = |\langle\psi(t)|\psi_-(t)\rangle|^2, \qquad (4.16)$$

来判断每一时刻是否能够绝热跟随, 其中 $|\psi(t)\rangle$ 是系统的真实态. 如果系统完全跟随绝热基态演化 (绝热跟随), 瞬时保真度的值应该是 1. 显然, 在 $t = t_{\text{fin}}$ 时, $F(t) = F_{\text{fin}}$. 不同 λ 值, $\alpha = 0.5, v_0 = 5$ 时, $F(t)$ 随时间的变化如图 4.6 所示. 很明显, $\lambda = 0$ 时, 超绝热过程实现了, 在所有时刻, 保真度的值都保持在 1.

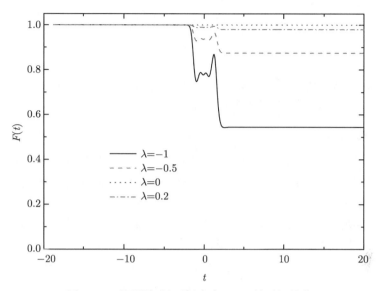

图 4.6　不同附加场下保真度 $F(t)$ 的时间演化

4.1.3　Demkov-Kunike 模型的高保真度超绝热量子驱动

本节采用上述方法讨论 Demkov-Kunike 模型的高保真度超绝热量子驱动问题[44]. Demkov-Kunike 模型中 $\gamma(t)$ 和 $v(t)$ 的形式如下[54,55]：

$$v(t) = v_0\mathrm{sech}(t/T), \quad \gamma(t) = \gamma_1 + \gamma_0 \tanh(t/T), \tag{4.17}$$

其中, T 是外场的扫描周期, 也称脉冲的宽度; γ_1 是静态失谐参数; γ_0 是啁啾参数; v_0 是最大耦合参数. 假设初始时刻 $t = t_{\mathrm{ini}}$ 系统制备在绝热基态 $|\psi_-(t_{\mathrm{ini}})\rangle$ 上, 演化结束后终态 $|\psi_{\mathrm{fin}}\rangle$ 依然在绝热基态 $|\psi_-(t_{\mathrm{fin}})\rangle$ 上, 系统就能实现高保真的量子演化. 其保真度定义为[30]

$$F_{\mathrm{fin}} = |\langle \psi_{\mathrm{fin}} | \psi_-(t_{\mathrm{fin}}) \rangle|^2, \tag{4.18}$$

$F_{\mathrm{fin}} = 1$ 意味着高保真度演化过程. 对于 Demkov-Kunike 模型,

$$\begin{aligned}\dot{\theta} &= \frac{\dot{v}(t)\gamma(t) - v(t)\dot{\gamma}(t)}{T[v^2(t) + \gamma^2(t)]} \\ &= -\frac{v(t)\left[\gamma_0 + \gamma_1 \tanh\left(\dfrac{t}{T}\right)\right]}{T[v^2(t) + \gamma^2(t)]},\end{aligned} \tag{4.19}$$

不同于 4.1.2 小节, 取驱动总哈密顿量为

$$H_{\text{tot}}(t) = H + \lambda H_{\text{cd}}$$
$$= \gamma(t)\hat{\sigma}_z + v(t)\hat{\sigma}_x + \lambda\dot{\theta}\hat{\sigma}_y, \tag{4.20}$$

其中, λ 为附加场轻度参数. 可以发现附加场的大小依赖于能级差 $\gamma(t)$ 耦合强度 $v(t)$ 以及扫描周期 T. 在计算中用 T 作为标度, 设 $T = 1$.

1. 无静态失谐

$\gamma_1 = 0$, 意味着无静态失谐, Demkov-Kunike 模型简化为 Allen-Eberly 模型. 式 (4.19) 也被简化为 $\dot{\theta} = -v(t)\gamma_0/T[v^2(t)+\gamma^2(t)]$. 图 4.7 展示了不同附加场强度参数下等效能级差和等效的耦合强度随时间的变化, 图中参数分别为 $v_0 = 2$, $\gamma_0 = 0.1$, $T = 1$. 当附加场强度参数 $\lambda = 0$ 时, 对应两能级系统的 Allen-Eberly 模型. λ 不同, 等效能级差和耦合强度发生明显的变化.

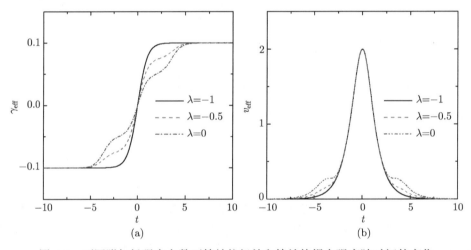

图 4.7 不同附加场强度参数下等效能级差和等效的耦合强度随时间的变化

图 4.8 展示了在不同附加场强度参数 λ 和啁啾参数 γ_0 下终态保真度 F_{fin} 随最大耦合强度 v_0 的变化规律. 当 $\lambda = 0$ 时, 对于小的耦合强度 v_0, 保真度 F_{fin} 随着 v_0 的增大单调递增. 对较大的 v_0, 出现了拉比型振荡. 随着 λ 的增大, 保真度明显提高, 直到 $\lambda = 1$ 时, 振荡消失, 系统获得超高的保真度.

图 4.9 展示了在不同附加场强度参数 λ 和耦合强度 v_0 下, 终态保真度 F_{fin} 随啁啾参数 γ_0 的变化规律. 当 $\lambda = 0$ 时, 不同范围的啁啾参数 γ_0 和耦合强度 v_0 对末态保真度 F_{fin} 的影响是不同的. 对于较小的啁啾参数 γ_0, 终态保真度 F_{fin} 主要依赖于耦合强度 v_0; 对于较大的啁啾参数 γ_0, 终态保真度 F_{fin} 随着啁啾参数 γ_0 的增大而逐渐降低. 同样, 当 $\lambda = 1$ 时, 超高的保真度在所有的参数范围内实现.

4.1 高保真度超绝热量子驱动

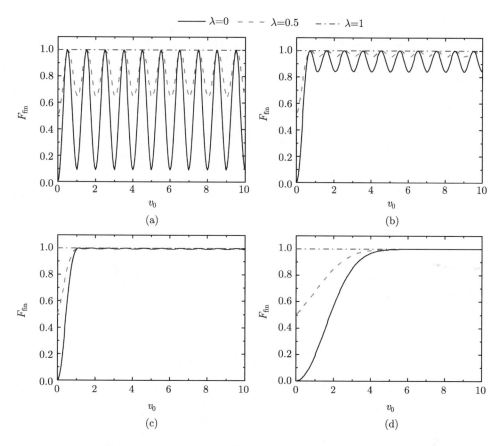

图 4.8 不同附加场强度参数和啁啾参数下末态保真度随最大耦合强度的变化规律
(a) $\gamma_0 = 0.1$; (b) $\gamma_0 = 0.5$; (c) $\gamma_0 = 1$; (d) $\gamma_0 = 15$

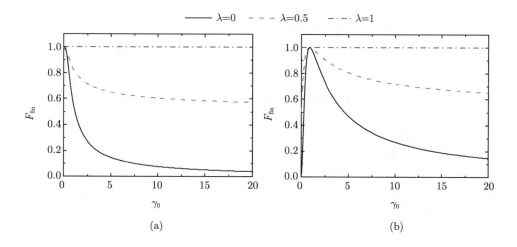

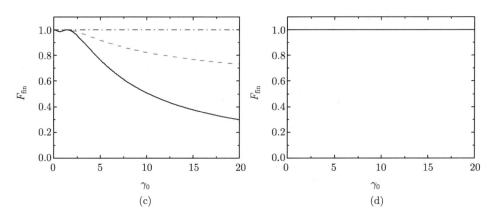

图 4.9　不同附加场强度参数和耦合强度下末态保真度随啁啾参数的变化规律

(a) $v_0 = 0.5$; (b) $v_0 = 1$; (c) $v_0 = 1.5$; (d) $v_0 = 10.5$

图 4.10 展示了不同啁啾参数 γ_0 下终态保真度 F_{fin} 关于附加场强度参数 λ 的变化规律, 可以看到保真度 F_{fin} 随着 λ 先增大后减小, 在 $\lambda = 1$ 时保真度 F_{fin} 达到最大值, 并且关于 $\lambda = 1$ 对称.

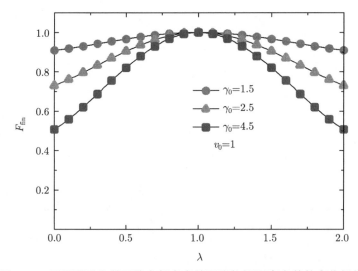

图 4.10　不同啁啾参数下终态保真度关于附加场强度参数的变化规律

为了展示超绝热技术的参数鲁棒性, 图 4.11 给出了在不同附加场强度参数下终态保真度关于啁啾参数和耦合强度的等密度图. 可以发现 $\lambda = 1$ 时, 振荡完全消失, 随着 γ_0 和 v_0 取值的不断变化, 系统的保真度几乎不变, 在所有的参数范围里实现了高保真度 [图 4.11(c)], 这个误差低于 10^{-4} [8]; 同时也表明了超绝热技术有很好的参数鲁棒性.

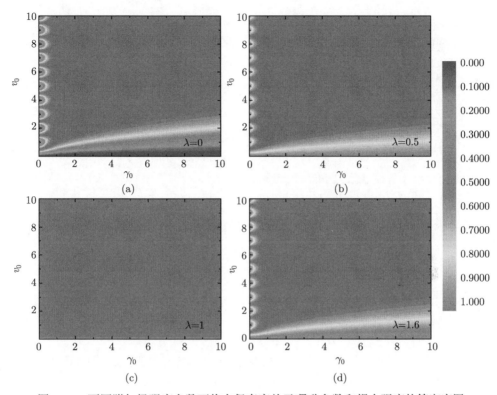

图 4.11 不同附加场强度参数下终态保真度关于啁啾参数和耦合强度的等密度图

2. 存在静态失谐情况

在 $\gamma_1 \neq 0$, 即存在静态失谐的情况下, 它会如何影响超绝热过程呢?

图 4.12 绘制 $\lambda = 1$ 时不同的静态失谐参数 γ_1 下的等效耦合强度 v_{eff} 和等效能级差 γ_{eff} 随时间 t 的变化图, 图中参数分别为 $v_0 = 2$, $\gamma_0 = 0.1$, $T = 1$. 可以看到, 不同的静态失谐值对等效耦合强度和等效能级差有不同的影响. 这是否会影响到终态保真度呢? 值得进一步讨论.

图 4.13 给出了不同附加场强度参数 λ 和最大耦合参数 v_0 下终态保真度 F_{fin} 随静态失谐参数 γ_1 的变化规律. 当 $\lambda = 0$ 时, 对于较大的 $|\gamma_1|$ 值, 终态保真度 F_{fin} 不依赖于 v_0 稳定在一个高保真度值; 对于小的 $|\gamma_1|$ 值, 保证度依赖于 v_0. 随着 λ 的出现, 保真度发生了明显的变化, 直到 $\lambda = 1$, 超高的保真度在所有的参数范围内实现. 更重要的是可以看到终态保真度关于 $\gamma_1 = 1$ 对称.

图 4.14 则给出了在静态失谐参数为负值 ($\gamma_1 < 0$) 时, 不同附加场强度参数 λ 和静态失谐 γ_1 下终态保真度 F_{fin} 关于耦合强度 v_0 的变化规律. 当 $\lambda = 0$ 时, 对于较小的 $|\gamma_1|$, 保真度 F_{fin} 随着 v_0 的增大单调递增, 并且出现了拉比型振荡

[图 4.14(a)]; 当 $|\gamma_1| = 0.1$ 时, 保真度 F_{fin} 随着 v_0 的增大而增大, 保持在一个较大的值 [图 4.14(b)]. 对于较大的 $|\gamma_1|$, 保真度 F_{fin} 随着 v_0 的增大而减小, 再次出现一个振荡 [图 4.14(c)], 并且随着 $|\gamma_1|$ 的继续增大, 保真度 F_{fin} 稳定在超高值 [图 4.14(d)]. 随着 λ 出现, 保真度 F_{fin} 明显提高, 直到 $\lambda = 1$ 时, 振荡完全消失, 保真度实现超高值.

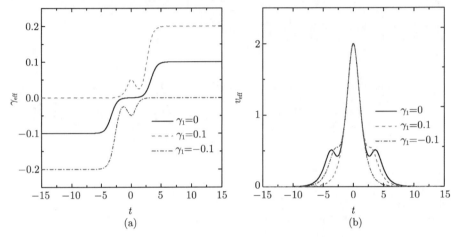

图 4.12 $\lambda = 1$ 时不同的静态失谐参数下的等效耦合强度和等效能级差图

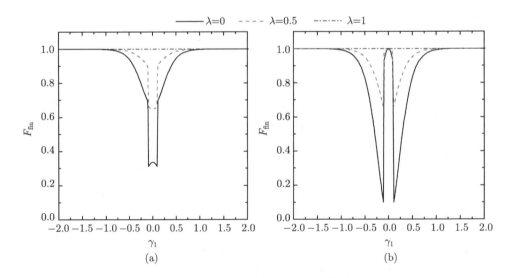

图 4.13 不同附加场强度参数和最大耦合参数下终态保真度随静态失谐参数的变化规律

(a) $v_0 = 0.2$; (b) $v_0 = 1.5$

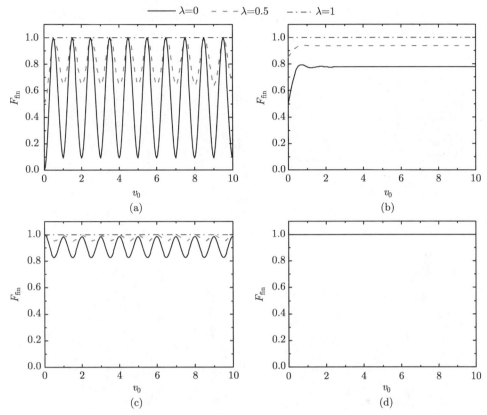

图 4.14 不同附加场强度参数和静态失谐下终态保真度随最大耦合强度的变化规律

(a) $\gamma_1 = -0.01$; (b) $\gamma_1 = -0.1$; (c) $\gamma_1 = -0.5$; (d) $\gamma_1 = -2$

在存在静态失谐情况下，超绝热技术也具有参数鲁棒性，给出了在不同附加场强度参数下终态保真度关于静态失谐参数和耦合强度的等密度图 (图 4.15)。可以发现 $\lambda = 1$ 时，振荡完全消失，随着 γ_1 和 v_0 取值的不断变化，系统的保真度几乎不变，在所有的参数范围里实现了高保真度 [图 4.15(c)]。同样，从图 4.15 中看到，系统的保真度关于 $\gamma_1 = 0$ 对称，这是由于两能级系统中 Demkov-Kunike 模型跃迁概率为[55,56]

$$P = \frac{\cosh(2\pi T \gamma_0) - \cosh\left[2\pi T (\gamma_0^2 - v_0^2)^{\frac{1}{2}}\right]}{2\cosh(\pi T \gamma_1) + \cosh(2\pi T \gamma_0)}, \tag{4.21}$$

其中，P 是系统的跃迁概率，很明显 P 是关于 $\gamma_1 = 0$ 对称的。进一步分析也同样发现，在存在静态失谐的情况下，系统的终态保真度 F_{fin} 关于 λ 具有对称性。

为了进一步检验超绝热技术的有效性，定义系统瞬时保真度 $F(t)$，其形式为[43]

$$F(t) = |\langle \psi_t | \psi_-(t) \rangle|^2, \tag{4.22}$$

其中，$|\psi_t\rangle$ 是系统演化过程中的真实态. 图 4.16 展示了不同附加场强度参数下保真度随时间变化规律. 可以发现无论取任何参数，只要 $\lambda=1$，系统在演化过程中的每一时刻都能实现绝热跟随，并且具有很高的保真度.

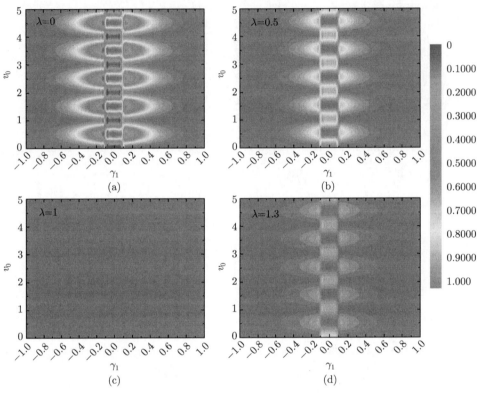

图 4.15　不同附加场强度参数下终态保真度关于静态失谐参数和耦合强度的等密度图

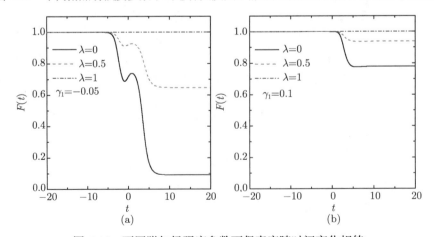

图 4.16　不同附加场强度参数下保真度随时间变化规律

4.1.4 结论与讨论

本节主要研究了两能级系统高保真度超绝热量子驱动问题,以啁啾 Gaussian 模型和 Demkov-Kunike 模型为例,展示了超绝热技术的优越性. 该方法能有效抑制跃迁概率的非绝热振荡,从而大大提高了演化过程的保真度. 合适的附加场能确保系统在短时间内绝热跟随,期望该方法在实验上能够实现. 例如,在加速光晶格玻色–爱因斯坦凝聚中,时间依赖的线性失谐可以通过准动量来控制,而耦合则可以通过晶格激光束来实现[30];或者应用在波导耦合器系统中,在该系统中 L 表示波导管长度,γ 表示传播常数之差,可以通过改变波导管的横截面积调节,v 表示耦合参数,可以通过改变波导管之间的距离调节[22]. 该方法由于具有高保真度、好的参数鲁棒性以及快速性,在量子信息科学 (如量子计算、量子通信以及量子度量学等) 领域中成为一个非常重要的工具[5,57].

4.2 高保真度超快量子驱动

4.1 节研究了两能级系统超绝热量子驱动问题,本节将讨论非线性两能级系统中高保真度超快量子驱动问题. 主要包括两个方面:一是具有粒子间相互作用的非线性两能级系统;二是具有粒子间相互作用和非线性扫描两种非线性的两能级系统,探讨了超快量子驱动的一些规律.

4.2.1 具有粒子间相互作用两能级系统中的超快量子驱动

量子系统的精确调控是现代科学领域的一个基本要求[3],它涉及量子信息[58]、分子系统的相干操控[59-61] 以及精密测量[4,62,63] 等. 量子操控的最终目标就是制备一个人们需要的高保真度量子态.

当人们在应用量子信息科学[58],如量子计算、量子通信以及量子计量时,保真度的要求是十分严格的,原则上人们的目标是具有极完美的保真度. 在量子调控中,最著名的应用也许就是量子计算,人们要求不仅要精确控制,同时其参数变化还要在大数量独立量子比特 (qubits) 和重复的量子门操作中具有高的稳定性. 事实上,随着量子比特数和量子门操作的增加,误差发生的可能性也随之增大. 量子信息编码仅在误差远低于一个临界阈值时才得到有效保护,一般情况,这个误差阈值非常严格,它依赖于误差修正技术和量子比特序列假定,其范围通常在 $10^{-4} \sim 10^{-2}$ [57],这就要求过程中要有高保真度. 另外,在量子信息等过程中,量子退相干 (quantum decoherence)[64] 是一个基本的阻碍,当人们在应用量子信息时,量子退相干不约而同地成为要考虑解决的主要问题[57],量子控制过程要尽可能比典型的退相干时间快,同时还要确保量子控制技术本身不产生大的误差,对参数具有鲁棒性. 因此,为了使退相干影响最小,同时具有较大的稳定性和参数鲁棒性,通常要求在尽可能短的时间内以高保真度实现量子操控,从而使量子操控具有精确性、持久性、高保真性和快速性。

近几年,量子速度极限 (quantum speed limit) [65] 优化控制理论已被建立,特别是 Bason 等在光晶格 BECs 的线性两能级系统中实现了高保真度超快量子驱动 (high-fidelity super-fast quantum driving) [30]。所考虑的系统为一个简单的 Landau-Zener (LZ) 模型,其哈密顿量如下:

$$H(t) = \Gamma(\tau)\hat{\sigma}_z + \omega(\tau)\hat{\sigma}_x, \tag{4.23}$$

其中,$\hat{\sigma}_x$ 和 $\hat{\sigma}_z$ 是泡利算符;$\Gamma(\tau)$ 和 $\omega(\tau)$ 分别是两个非绝热态 $|0\rangle$ 和 $|1\rangle$ 的能级差和耦合强度,$\tau = t/T \in [0,1]$ 是重新尺度化的时间,这里 ω 为常数,$\Gamma(\tau)$ 随时间变化并且 $\Gamma(1) = -\Gamma(0) = 2$。系统的瞬时绝热能级为 $|\psi_{g,e}(\tau)\rangle$,能隙为 2ω,系统在时间 T 内从初始绝热基态 $|\psi_{\text{ini}}(0)\rangle = |\psi_g(0)\rangle$ 到终态 $|\psi_{\text{fin}}(1)\rangle$ 演化。他们的目标是设计一个调控技术,以驱动量子系统并以如下的一种方式经过反交叉点:在演化结束 $\tau = 1$ 时,终态 $|\psi_{\text{fin}}\rangle$ 尽可能在绝热基态 $|\psi_g(1)\rangle$ 上,并且具有保真度 $F_{\text{fin}} = 1$。这里保真度 F_{fin} 定义如下:

$$F_{\text{fin}} = |\langle \psi_{\text{fin}}(1)|\psi_g(1)\rangle|^2. \tag{4.24}$$

在 Hilbert 空间中,应该有无数种路径连接初态 $|\psi_g(0)\rangle$ 和终态 $|\psi_g(1)\rangle$,首先考虑了一种特殊情况:演化过程中用最小的时间 T,实现最终的量子力学速度极限,后来人们称之为超快量子控制技术 (super-fast quantum control protocol) [57]。其两能级系统和实验实现如图 4.17 所示 (也见图 2.4)。图 4.17(a) 表示 LZ 两能级系统,其中 E_{rec} 是系统的自然能量尺度,非绝热态 (裸态) $|0\rangle$ 和 $|1\rangle$ 被耦合到绝热态 $|\psi_g\rangle$ 和 $|\psi_e\rangle$,非绝热能级在 $\tau = 0.5, E = 0$ 处交叉,箭头表示两种极端技术;图 4.17(b) 表示在 BEC 中的实验实现,凝聚体囚禁在光偶极阱中并装载在光晶格里;而图 4.17(c) 表示凝聚体在光晶格中的能带结构,盒子表示用来实现在最低的两个能带中等效两能级系统的准动量范围。

实验中凝聚体囚禁在光偶极阱的光晶格里,采用了一种超快 (时间最小) 的复合脉冲 (CP) 技术,并且与通常的线性 LZ 绝热控制技术和局部的绝热 Roland-Cerf (RC) 技术做了比较,各种驱动技术如图 4.18(a)~(c) 所示,其耦合强度均为常数,而能级差分别为

$$\Gamma(\tau) = 0, \tau \in (0,1), \quad \text{CP} \tag{4.25}$$

$$\Gamma(\tau) = 4E_{\rm rec}(\tau - 0.5), \quad {\rm LZ} \tag{4.26}$$

$$\Gamma(\tau) = \frac{4\epsilon\omega^2 T(\epsilon)(\tau - 1/2)}{\sqrt{1 - 16\epsilon^2\omega^2 T^2(\epsilon)(\tau - 1/2)^2}}, \quad {\rm RC} \tag{4.27}$$

其中, 式 (4.27) 中总时间 $T(\epsilon) = 1/(\epsilon\omega\sqrt{4+\omega^2})$, ϵ 为一个小量.

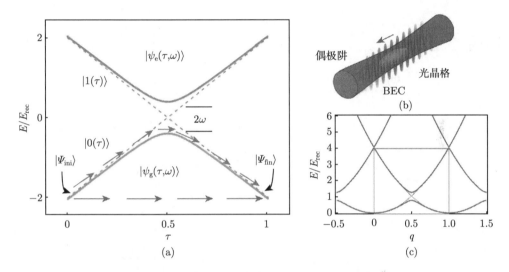

图 4.17　两能级量子系统示意图和实验实现 [30]

(a) LZ 两能级系统; (b) 在 BEC 中的实验实现; (c) 凝聚体在光晶格中的能带结构

以实现保真度 0.9 为标准, 发现复合脉冲技术所需时间最少, 几乎近似于速度极限, 而 RC 技术实现同样目标态的时间是 CP 技术的 2 倍. 实验和理论模拟结果如图 4.18(d) 和 (e) 所示, 其中 RC (圆), LZ(方块), CP (三角), 虚线表示理论预言, 4.18(e) 中水平点线表示 0.9 的保真度阈值, 所有的实验中 $\omega = 0.5$.

然而, 这些研究仅限于线性系统, 其中粒子之间的相互作用等非线性项被忽略了. 事实上, 在 BECs 中, 原子密度相对较大, 粒子间的相互作用不能被忽略.

本节研究非线性两能级系统中的超快量子驱动, 其中, 两能级系统的能级依赖于布居数, 为粒子间平均场相互作用, 考察了粒子间相互作用对超快量子驱动的影响. 研究表明, 粒子间相互作用倾向于增大给定初始态到终态时实现高保真度的最小时间, 并且当超过某个临界值时, 这个最小时间变成无穷大, 说明要实现高保真度已不可能了 [66].

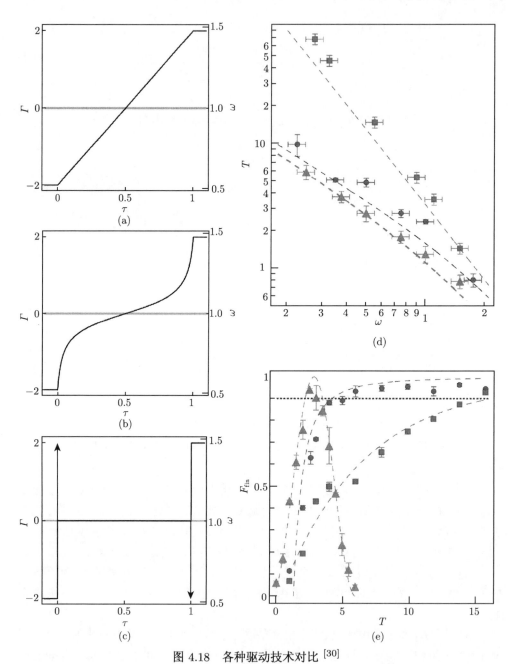

图 4.18 各种驱动技术对比 [30]

(a) LZ; (b) RC; (c) CP; (d) CP 技术中实现优化保真度 $F_{\text{fin}} = 0.9$ 所需要的时间; (e) 终态保真度随持续时间 T 的变化情况

4.2 高保真度超快量子驱动

1. 非线性超快量子驱动

考虑的模型仍然和第 3 章相同, 由无量纲化的非线性 Schrödinger 方程描述:

$$\mathrm{i}\frac{\partial}{\partial t}\begin{pmatrix} a \\ b \end{pmatrix} = H(t)\begin{pmatrix} a \\ b \end{pmatrix}, \tag{4.28}$$

其中哈密顿量为

$$H(t) = \left[\varGamma(\tau) + c(|b|^2 - |a|^2)\right]\hat{\sigma}_z + \omega(\tau)\hat{\sigma}_x, \tag{4.29}$$

其中, a 和 b 分别为非绝热态 $|0\rangle$ 和 $|1\rangle$ 的概率幅; c 为非线性参数, 表示粒子间相互作用.

本节研究具有粒子间相互作用的非线性系统的超快量子驱动, 考察非线性相互作用将怎样影响量子速度极限. 线性系统中, 通过运用 CP 技术, 初始量子态能在最短的允许时间内到达终态, 实现最大的速度. 在非线性两能级系统中, 由于出现了非线性, 隧穿动力学发生了很大的变化, 此时, Schrödinger 方程 (4.28) 不能够解析求解, 同样采用变步 4-5 阶 Runge-Kutta 去数值跟踪量子演化并计算保真度和到达预想终态的最小时间. 为了和文献 [30] 比较, 在所有的数值计算中, 取耦合强度 $\omega = 0.5$.

图 4.19(a) 显示了在不同非线性相互作用下 CP 技术对应的终态保真度, 从图中可以看到, 高保真度到达预想终态的最小时间依赖于非线性相互作用, 对于弱的相互作用, 保真度在 1 和 −1 之间对称地振荡; 随着非线性的加强, 振荡周期增大, 即最小时间也随着增大; 更为有趣的是, 当非线性超过一个临界值时, 尽管保真度的振荡仍然存在, 但是其最大幅度却不能到达 1, 这意味着存在一个非线性相互作用的临界值, 超过此值时, 高保真度的超快量子驱动将不再实现.

为了评估 CP 脉冲技术的优越性, 也将它和非线性 LZ 技术以及局部绝热的 RC 方法做了对比. 计算结果如图 4.19 所示, 水平虚线表示保真度为 0.9 的阈值. 很明显, 对于弱非线性粒子间相互作用, 非线性倾向于增加初态到终态的最小时间, 其中 LZ 技术比 CP 技术更慢, 而 RC 方法到达保真度 $F_{\text{fin}} = 0.9$ 的时间介于 CP 技术 和 LZ 技术之间. 相似于 CP 技术, 当非线性相互作用超过一个临界值时, 对于 LZ 技术 和 RC 方法, 高保真度超快量子控制也不能实现.

以上过程可以通过分析非线性系统的演化加以理解, 图 4.20 显示了到达保真度 $F_{\text{fin}} = 0.9$ 的最小时间随非线性相互作用的变化情况. 由子图中弱相互作用范围时 LZ (方块) 和 RC (三角) 方法所使用的最小时间 $T_{0.9}$ 可以看到: ① CP 技术中到达终态的速度最快; ② 不论 CP 技术, 还是 LZ 技术或 RC 方法, 由于非线性作用, 到达终态的最小时间将增大; ③ 存在一个临界值, 当非线性超过这个值时, 最

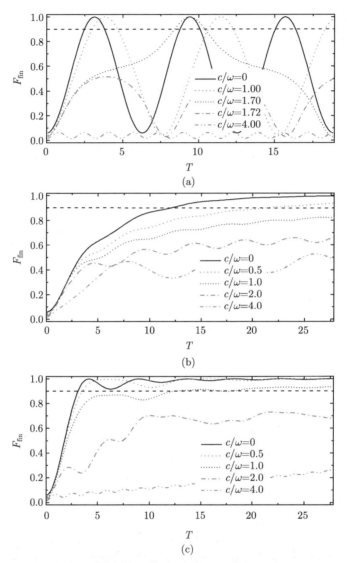

图 4.19 不同非线性相互作用下终态保真度随持续时间的变化情况

(a) CP 技术; (b) LZ 技术; (c) RC 方法

小时间会发散. 例如, 对 CP 技术, 其临界值是 $c = 0.8522$, 即 $c/\omega = 1.7044$. 该现象可以通过分析非线性系统的演化来很好地理解 [67], 对于弱非线性相互作用, 能观察到 Josephson 振荡, 对于大的非线性相互作用, 系统会出现自俘获现象. 对于非线性 LZ 技术和 RC 方法, 非线性倾向于增加从基态到激发态的隧穿概率. 在线性情况下, 系统初始制备在绝热基态上并以有限的概率隧穿到激发态, 导致其 LZ 技术和 RC 方法的保真度分别为 $F_{\text{fin}} = 1 - \exp(-\pi T \omega^2 / 4)$ 和 $F(\tau) \geqslant 1 - \epsilon^2$, 只有在绝热极限下, 保真度 $F_{\text{fin}} = 1$. 然而, 实现该绝热性的时间变为无穷. 在弱非线性相互作用下, LZ 隧穿概率随扫描率 $1/T$ 的变化在线性时的指数律会有一个非线性修正 [68], 并且修正因子随着非线性的增加单调减小, 这意味着隧穿概率增加. 因此, 随着非线性的加强, 保真度减小, 从而导致到达 $F_{\text{fin}} = 0.9$ 的最小时间增加. 当非线性参数超过临界值时, 即使在绝热极限下, LZ 隧穿概率也不会为零 [68-71], 因此高保真度到达终态的最小时间会发散, 为无穷大.

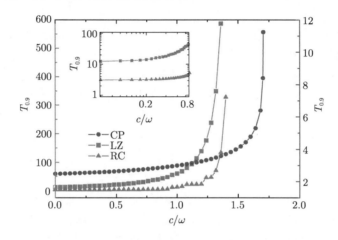

图 4.20 实现保真度 $F_{\text{fin}} = 0.9$ 所需的最小时间随非线性相互作用的变化情况

为了进一步理解这种临界行为, 定义了一个最大保真度 $F_{\max}$, 用来表示持续时间 T 从 0 到 ∞ 演化过程中终态保真度的最大值, 计算结果如图 4.21 所示.

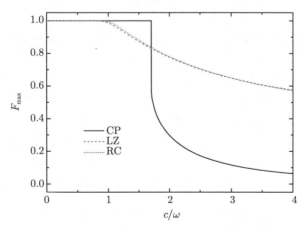

图 4.21 最大保真度随非线性相互作用的变化情况

由图可见，$F_{\max}$ 和非线性相互作用密切相关，对于弱相互作用，$F_{\max}$ 的值是 1，当非线性超过临界值后，对于 CP 技术，它很快降低，临界值是 $c/\omega = 1.7044$，然而对于 LZ 技术和 RC 方法，临界值 c/ω 大约为 1，当 $c/\omega > 1$ 后，最大保真度相对于 CP 技术而言缓慢降低. 需要注意的是，在 LZ 技术和 RC 方法中，实现完全绝热，即 $F_{\max} = 1$ 所需要的时间均趋于无穷.

2. 实验实现与结论

以上讨论的非线性两能级系统中的超快量子驱动可以在加速光晶格 BECs 的两个 Bloch 带间实现，其中高密度的原子可以存在，另外，可以利用 Feshbach 共振来调节散射长度，以至于可以观察到非线性效应. 其数学模型可以用式 (4.29) 来描述，其中 $\Gamma(\tau)$ 和 $\omega(\tau)$ 能通过准动量 q 和光晶格深度 V_0 来控制. 系统初始制备在晶格 $q = 0$ (对应于 F_{ini}) 的低能带上，经过一段时间 T 的演化到达目标态 F_{fin}，BECs 可以加载在 $V = (V_0/2)\cos(2\pi x/d_{\text{L}})$ 形式的光晶格势上，其中 d_{L} 表示晶格空间. 非线性 $c/\omega = 8\pi\hbar^2 n_0 a_s/mV_0$ [72]，其中 n_0 是凝聚体的平均密度，m 是原子质量，a_s 是 s 波散射长度. 在一些典型的实验中 [30]，取 $\omega = 0.5$ 对应于 $V_0 = 2E_{\text{rec}} = 2\hbar\omega_{\text{rec}}(\omega_{\text{rec}} = 2\pi \times 3.15\text{kHz})$，对铷原子 $a_s = 5.4\text{nm}$ [73]，这样当 $n_0 = 1 \times 10^{19} \sim 1 \times 10^{21}\text{m}^{-3}$ 时，$c/\omega = 0.025 \sim 2.5$，高的原子密度意味着大的非线性 c. 另外，人们可以通过磁场 Feshbach 共振来对粒子间相互作用 c 任意调节. 在文献 [30] 的实验中，参数 $c \leqslant 0.05$，因此相互作用被忽略了 [31]，但是，导致实验和理论值之间存在一定的偏差. 本节理论正好解释了这种现象，比率 c/ω 越大，其非线性效应就越明显. 值得注意的是，近些年，已经在具有周期势的 BECs 的两能带之间观察到非线性 LZ 隧穿 [71,74]，这说明非线性两能级系统中的超快量子驱动也可以在实验上实现.

总之，本节研究了非线性两能级系统的高保真度超快量子驱动问题，考察了原子间相互作用对高保真度量子驱动的影响. 与线性情况相似，CP 技术能实现超快量子驱动，即对比于非线性 LZ 和 RC 技术，它实现高保真度所需要的最小时间最短. 经研究发现粒子间相互作用会增加这个最小时间，同时存在一个临界值，当非线性超过这个临界值后，实现高保真度量子驱动几乎不可能，其最小时间会变为无穷. 尽管探讨的实验聚焦于加速光晶格中 BECs 的等效两能级系统，但是结果具有普适性，可以推广到任意一个两能级系统中，不仅可以在 Bloch 带间，而且也可以到任意外场的布居数翻转问题以及耦合势表面的凝聚运动等物理系统中 [66,70].

4.2.2 广义非线性两能级系统中高保真度超快量子驱动

本节将超快量子驱动问题推广到广义非线性两能级系统中. 该系统中包含两种非线性: 粒子间相互作用和非线性能级扫描. 研究表明，超快量子动力学和粒子

4.2 高保真度超快量子驱动

间相互作用、能级扫描参数以及能级扫描强度都有关系. 大的能级扫描参数能在更短的时间内实现高保真度的量子驱动, 而粒子间相互作用会使线性系统中的量子速度极限遭到破坏. 对于吸引相互作用, 到达目标态的最小时间随着能级扫描强度的增加而单调增大, 然而对于排斥相互作用, 反而随着扫描强度的增大而减小. 通过临界行为的讨论, 获得了临界相互作用或临界扫描强度的解析表达式, 发现当相互作用或者扫描强度超过一个临界值时, 高保真度量子驱动不能实现.

量子系统的精确控制是一个既重要又富有挑战性的问题, 在许多领域都有着重要的应用[1,3,4,62], 如在量子信息[5,58]、量子光学[75]、玻色–爱因斯坦凝聚、核磁共振以及十分普遍的原子分子及化学物理中[59-61]. 速度、保真度和鲁棒性是量子控制技术中的三个关键因素[15]. 源于 Heisenberg 的时间–能量不确定原理, 在量子力学中, 量子系统的演化过程有一个基本限制: 最大速度[76]. 量子速度极限 (quantum speed limits, QSLs) 回答了量子系统在演化过程中应该快到什么程度, 并且给出了量子演化速度的理论上限[19,77-83]. 事实上, 量子系统的操控过程中, 为了最小化量子退相干的作用, 不仅需要快的速度, 即最少的演化时间[57,64], 而且需要操控技术具有高保真度和参数鲁棒性. 为了实现这一目标, 一系列优化控制技术和加速方法相继被提出[65,84], 如量子最速下降 (quantum brachistochrone)[85]、超绝热 (superadiabatic) (也称为 transitionless, counterdiabatic) 技术量子驱动[11,13]、绝热捷径方法 (shortcut to adiabaticity)[14,18] 以及缀饰态方法[86].

然而, 大部分的实验和相关的理论工作都限制于线性两能级系统[30,32]. 事实上, 真实的物理系统是一个非线性量子系统. 一般而言, 基于最初的线性两能级系统, 在非线性方面有两个方面的扩展: 第一种情况是在线性两能级系统的基础上增加粒子间相互作用的非线性项[72,87], 用来描述一个时间依赖的两态系统玻色–爱因斯坦凝聚的平均场近似; 第二种情况是能级是一个非线性的扫描外场, 可以以任意设想的方式操控量子系统[42,88]. 这些非线性能极大地影响系统的跃迁动力学[42,69,72,87,88], 特别在超冷玻色–爱因斯坦凝聚体系中. 在 LZ 跃迁中, 人们已经研究了具有粒子间相互作用和非线性扫描的组合效应, 发现了许多有趣的现象[41]. 对于高保真度的量子驱动, 仅在近几年, 人们才去处理复杂的非线性量子系统[89-91]. 高保真度的超快和超绝热量子驱动也已经在非线性能级扫描的广义 LZ 模型的实验中被实现[31]. 对于包含粒子间相互作用的非线性两能级系统, 也分别在 LZ、RC 和 CP 模型中研究了高保真度的超快量子跃迁动力学[66]. 然而, 以上关于超快量子驱动的研究, 要么仅考虑了粒子间相互作用, 要么只考虑了非线性扫描[31,66]. 因此, 探索同时具有粒子间相互作用和非线性扫描两种非线性的量子系统中的超快跃迁就显得非常有意义, 会出现哪些有趣的现象呢?

本节将探讨以上同时具有粒子间相互作用和非线性扫描两种非线性的两能级系统中的超快量子驱动问题.

1. 模型

非线性两能级模型由以下无量纲化的 Schrödinger 方程来描述 [72]:

$$i\frac{\partial}{\partial t}\begin{pmatrix} a \\ b \end{pmatrix} = H(t)\begin{pmatrix} a \\ b \end{pmatrix}, \tag{4.30}$$

其中哈密顿量为

$$H(t) = \left[\Gamma(t) + c(|b|^2 - |a|^2)\right]\hat{\sigma}_z + \omega(t)\hat{\sigma}_x. \tag{4.31}$$

其中, a 和 b 分别表示非线性态 $|0\rangle$ 和 $|1\rangle$ 的概率幅; 总概率 $|a|^2 + |b|^2 = 1$ 守恒; $\hat{\sigma}_x$ 和 $\hat{\sigma}_z$ 为泡利矩阵; $\Gamma(t)$ 和 $\omega(t)$ 分别表示两非绝热能级之间的能级和耦合强度; 参数 c 表示粒子间相互作用. 在上述模型中, $c < 0$ 表示排斥相互作用, 而 $c > 0$ 表示吸引相互作用.

以上系统有瞬时绝热本征态 $|\psi_{g,e}(t)\rangle$, 其中下标 g 和 e 分表代表基态和激发态. 假设初始时 $(t = 0)$, 系统制备在绝热基态 $|\psi_g(0)\rangle$ 上. 经历演化时间 T 后, 终态在 $t = T$ 时为 $|\psi_{\text{fin}}\rangle$. 简单起见, 将时间重新尺度化, 用 $\tau = t/T, \tau \in [0, 1]$. 所考虑的模型中耦合是常数, 而能级差是时间的函数, 如下所示:

$$\omega(\tau) = \omega(\text{const}), \quad \Gamma(\tau) = \text{sgn}\left(\tau - \frac{1}{2}\right)\Gamma_0 \left|2\left(\tau - \frac{1}{2}\right)\right|^\beta, \tag{4.32}$$

其中, Γ_0 是能级扫描强度, 表示能级扫描范围为 $-\Gamma_0 \sim \Gamma_0$; β 控制能级的非线性, 称为幂律参数或者非线性能级扫描参数. 研究目标是采用量子控制技术 (4.32), 以尽可能短的时间驱动系统从初始态 $|\psi_{\text{ini}}\rangle = |\psi_g(0)\rangle$ 到终态 $|\psi_{\text{fin}}\rangle$. 同时, 在演化结束时 $(\tau = 1)$, 终态 $|\psi_{\text{fin}}\rangle$ 尽可能在绝热基态 $|\psi_g(1)\rangle$ 上, 实现高保真度 1. 这里保真度 F_{fin} 定义为

$$F_{\text{fin}} = |\langle\psi_{\text{fin}}|\psi_g(T)\rangle|^2, \tag{4.33}$$

用以特征化操控方法的效率.

显然, 不同的 $\Gamma(\tau)$ 和 $\omega(\tau)$ 对应于不同的操控技术. 当 $\beta = 1$ 时, 能级扫描是线性的, 对应于 LZ 方法, 而 $\beta \to \infty$ 则代表 CP 方法. 图 4.22 显示了不同能级扫描参数 β 时, $\Gamma(\tau)$ 随时间的演化情况, 图中取 $\Gamma_0 = 2$.

当 $c = 0$ 时, 三种特殊的情况, 即 LZ 模型 ($\beta = 1$)、RC 模型和 CP 模型 ($\beta \to \infty$) 的高保真度超快量子驱动已经在光晶格玻色-爱因斯坦凝聚两能级系统

4.2 高保真度超快量子驱动

中的实验实现了 [30]. 尽管人们发现 CP 能实现时间最优控制, 达到量子速度极限, 然而, 还缺乏必要的证明 [30]. 这里所说的量子速度极限时间为 [30,65]

$$T_{\text{qsl}} = \frac{\arccos|\langle\psi_{\text{fin}}|\psi_{\text{ini}}\rangle|}{\omega} \tag{4.34}$$

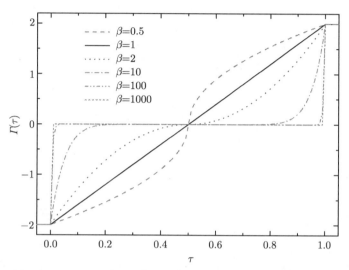

图 4.22　$\Gamma_0 = 2$ 时不同能级扫描参数 β 下 $\Gamma(\tau)$ 的时间演化

受文献 [30] 的启发, 实验上, 在非线性扫描的广义两能级系统中, 保持 $\Gamma_0 = 2$, 人们也实现了高保真度量子驱动, 发现要到达终态, 大的扫描参数 β 对应于更短的时间 [31]. 另外, 对于具有粒子间相互作用的非线性两能级系统 ($c \neq 0$), 在以上三种特殊模型时, 也初步研究了 $\Gamma_0 = 2$ 时的超快量子驱动过程, 发现粒子间相互作用能打破线性系统中的量子速度极限 [66].

2. 非线性超快量子驱动

本部分具体研究存在粒子间相互作用和非线性扫描两种非线性的量子系统中的超快量子驱动, 分析粒子间相互作用、能级扫描参数以及能级扫描强度等参数对跃迁动力学的影响. 由于出现了非线性, Schrödinger 方程 (4.30) 不再解析求解, 因而, 仍然采用 4-5 阶 Runge-Kutta 法进行数值求解.

通过取不同的参数 c 和 β, 可以计算系统 (4.30) 的动力学过程. 图 4.23 给出了终态保真度随总演化时间 T 的变化情况, 图中分别取 $\beta = 0.5$、1、10、100、1000, $c/\omega = -4, -1, 0, 1, 1.7, 1.72$, 其中 $\omega = 0.5, \Gamma_0 = 2$. 可以看到, 终态保真度存在一个振荡行为, 并且和粒子间相互作用、能级扫描参数密切相关. 在一些参数下, 保真度的最大振荡幅度能到达 1, 而在另一些参数下, 其值始终不能到达这个值.

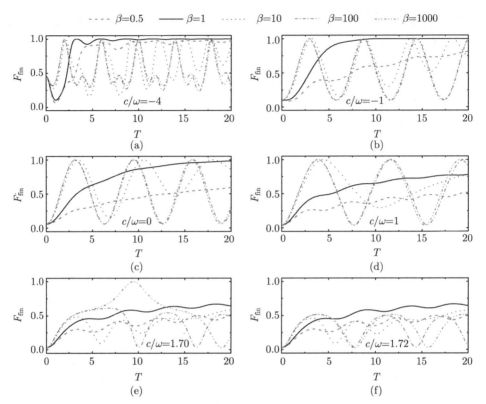

图 4.23　终态保真度在不同能级扫描参数和相互作用情况下随总演化时间的变化

更有意思的是，对于大的 β 值，终态保真度到达某一给定值所需的时间更短. 假设固定 $F_\text{fin}=0.9$，则到达这个保真度所需的最小时间 $T_{0.9}$ 也依赖于 β 和 c 的取值，如图 4.24 所示. 对于相同的粒子间相互作用值，到达具有高保真度终态的最小时间会随着能级扫描参数的增大而降低，且大的能级扫描参数值 β 对应的时间 $T_{0.9}$ 最短. 然而，对于同样的能级扫描参数值，到达终态的最小时间会随着相互作用从排斥到吸引逐渐增加，而且在非常强的吸引和排斥相互作用时，最小时间分别到达两个极限值. 不存在粒子间相互作用 (即 $c=0$) 时，对于足够大的 β，最小时间 $T_{0.9}$ 接近于量子速度极限. 对于足够强的吸引相互作用，最小时间会发散，而对于足够强的排斥相互作用，到达 $F_\text{fin}=0.9$ 的最小时间会是一个小值，甚至在相互作用很强时趋向于 0. 这就意味着高保真度的超快量子驱动在足够强的吸引相互下将不再实现，然而在足够强的排斥相互作用下，却能破坏线性系统中的量子速度极限. 这里需要注意的是，以上结论都是在保持能级扫描强度 $\varGamma_0=2$ 时所得到的.

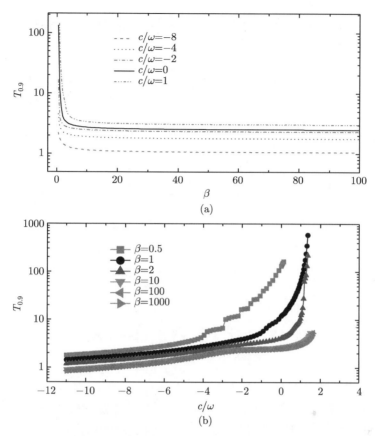

图 4.24 实现保真度 $F_{\text{fin}} = 0.9$ 的最小时间

(a) 在不同相互作用时随能级扫描参数 β 的变化; (b) 不同能级扫描参数下随相互作用 c 的变化

  以上研究发现, 到达高保真度终态的最小时间依赖于粒子间相互作用以及能级扫描参数. 那么, 它是否和能级扫描强度 $\varGamma_0$ 有关呢? 为此, 通过改变 $\varGamma_0$, 计算 $\beta = 1000$ 且保真度到达 $F_{\text{fin}} = 0.9999$ 时对应的最小时间 $T_{0.9999}$. 事实上, 当 $\beta = 1000$ 时, 对应的控制方法就是 CP 技术. 图 4.25 显示了不同相互作用下 $T_{0.9999}$ 随 $\varGamma_0$ 的变化情况. 很明显, 到达高保真度的最小时间也和扫描强度 $\varGamma_0$ 密切相关. 对于吸引相互作用, 最小时间 $T_{0.9999}$ 随着 $\varGamma_0$ 的增大而单调降低, 在足够大的 $\varGamma_0$ 时, 也能到达线性系统中的量子速度极限时间. 但是, 在排斥相互作用下, 最小时间却随着 $\varGamma_0$ 的增大而增加, 并且对于强的排斥相互作用, $T_{0.9999}$ 更小. 在很宽的 c、$\varGamma_0$ 参数范围, 线性系统中的量子速度极限被破坏, 甚至在非常强的排斥相互作用时, $T_{0.9999}$ 的值趋近于 0. 分析原因可以发现, 这是由于在这些参数范围存在 Josephson 振荡, 而且振荡周期随着非线性相互作用 c 越来越小 (排斥相互作用增加) 而减小, 随着 c 越来越大 (吸引相互作用增加) 而增大. 当相互作用足够大时, 对于一些特定

的 Γ_0，系统变成自俘获状态. 当临界行为出现时，高保真度量子驱动不再能够实现. 这时，$T_{0.9999}$ 变为发散的值.

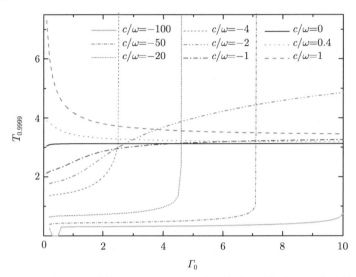

图 4.25 $\beta = 1000$ 时到达保真度 $F_{\text{fin}} = 0.9999$ 需要的最小时间在不同非线性相互作用下随能级扫描强度 Γ_0 的变化

为了进一步研究最小时间与非线性相互作用 c、能级扫描参数 β 以及能级扫描强度 Γ_0 的依赖关系，并理解临界行为，接下来考虑能级扫描参数值 β 很大的情况. 在某些情况下，如弱相互作用时，Josephson 振荡会出现，但是随着相互作用的加强并且超过一个临界值时，会发生自俘获现象. 因此，在发生自俘获现象的参数区域，高保真度超快量子控制就不再能够实现. 当自俘获发生时，系统的能量可以用经典哈密顿量来描述，即对于吸引相互作用 ($c > 0$)，$H_{\text{eff}}(c,\omega) = -c(1 - 2|a|^2)^2/2 - \omega\sqrt{1 - (1-2|a|^2)^2} < -\omega$，而对于排斥相互作用 ($c < 0$)，$H_{\text{eff}}(c,\omega) > \omega$[69,92]. 因此，可以获得发生临界行为的普适判据，即 $H_{\text{eff}}(c,\omega) < -\omega(c > 0)$ 或者 $H_{\text{eff}}(c,\omega) > \omega(c < 0)$. 这样，该判据可以表示如下：

$$\left(\frac{c}{\omega}\right)_{\text{cra}} > \frac{2\left[1 - \sqrt{1 - (1-2|a_i|^2)^2}\right]}{(1-2|a_i|^2)^2}, \quad c > 0, \tag{4.35}$$

$$\left(\frac{c}{\omega}\right)_{\text{crp}} < -\frac{2\left[1 + \sqrt{1 - (1-2|a_i|^2)^2}\right]}{(1-2|a_i|^2)^2}, \quad c < 0. \tag{4.36}$$

其中，$|a_i|^2$ 是对应的初态布居数. 当参数满足式 (4.35) 和式 (4.36) 时，超快量子驱动将不再实现. 根据式 (4.35)、式 (4.36)，也能获得实现高保真度量子控制的一些临界值，如当 $\Gamma_0 = 2$ 时，$(c/\omega)_{\text{cra}} = 1.7044$[图 4.23(a)]. 不同的 Γ_0 值对应于不同的临

4.2 高保真度超快量子驱动

界值 c/ω_{cra}. 同样, 也能获得在不同非线性相互作用时 Γ_0 的临界值, 如 $c/\omega = -4$ 时, $\Gamma_0 = 2.59808$. 图 4.26 中, 展示了实现高保真度时, c/ω 和 Γ_0 的参数范围. 其中阴影部分对应不能实现高保真度的参数区域, 而其他区域代表高保真度能够实现的参数区域. 圆点对应的参数值是 $(\Gamma_0, c/\omega) = (2.59808, -4)$. 显然, 在吸引相互作用下, 当 $c/\omega < c/\omega_{\text{cra}}$ 时, 高保真度量子驱动能实现, 而当 c/ω 超过临界值 c/ω_{cra} 时, 高保真度量子驱动将不再能够实现. 对于排斥相互作用, 当 $\Gamma_0 < 2.59808$ 时, 在所有相互作用参数范围, 高保真度都能实现, 然而, 当 $\Gamma_0 > 2.59808$ 时, 在某些排斥相互作用范围, 高保真度量子驱动却不再实现.

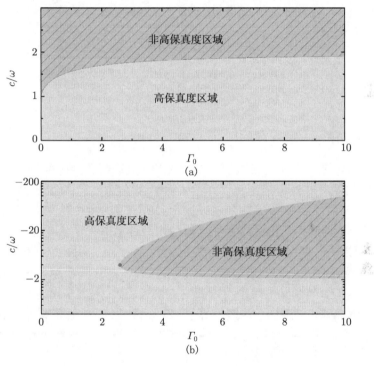

图 4.26 Γ_0 和 c/ω 空间的相图

(a) $c > 0$; (b) $c < 0$

为了更清晰地展现优化控制技术, 可以采用 Bloch 球来展示上述过程[91]. 在 Bloch 球上, 初态和终态位于球的北极和南极附近, 每种驱动技术对应于一条从初态到终态的路径. 图 4.27 显示了不同粒子间相互作用下, $\beta = 1$ (虚线), $\beta = 2$ (虚-点线) 和 $\beta = 1000$ (实线) 时不同控制技术的 Bloch 球表示: 图 4.27(a) $c/\omega = -2$, 其中另一条线表示 $\beta = 0.5$ 的情形; 图 4.27(b) $c/\omega = 0$; 图 4.27(c) $c/\omega = 1$. 五角星分别代表系统的初态和目标态. 图中 $\Gamma_0 = 2$. 很明显, 对于大的 β 值, 路径更短, 意味着量子控制速度更快 (即到达终态的最小时间更短). 对于小的 β 值, 如 $\beta = 0.5$ [图 4.27(a)], $\beta = 1$ [图 4.27(b) 和 (c)], 为了实现高保真度, 路径很长, 而且在 Bloch 球上不断缠绕, 特别在终态附近.

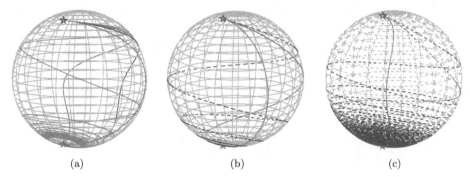

图 4.27　不同相互作用和扫描参数时对应不同路径的 Bloch 球表示

(a) $c/\omega = -2$; (b) $c/\omega = 0$; (c) $c/\omega = 1$

3. 结论与讨论

本节研究了非线性两能级系统中的超快量子驱动问题, 探索了粒子间相互作用和非线性扫描两种非线性在超快动力学中的组合效应. 一个重要的结果是, 在同样的相互作用值时, 大的扫描参数到达高保真度所需要的最小时间更短, 甚至在大的扫描强度下, 弱吸引相互作用或者没有相互作用时能到达量子速度极限. 粒子间排斥相互作用倾向于减少到达目标态的最小时间, 甚至在某些扫描强度参数范围趋近于 0, 从而使得线性系统中的量子速度极限遭到破坏. 然而, 不论是吸引还是排斥相互作用, 总存在一个临界相互作用或者临界扫描强度, 超过临界值时, 高保真度量子操控不能再实现, 也获得了这些临界值的解析表达式. 上述结果可以在一个加速光晶格玻色–爱因斯坦凝聚的两个 Bloch 带之间实现, 其数学模型用方程 (4.30) 描述, 其中 $\Gamma(\tau)$、ω 和 c 能分别通过准动量、晶格深度和原子密度或者 Feshbach 共振去调节[31,71,74,93]. 初始时刻, 系统制备在 $q = 0$ (对应于 $|\psi_{\text{ini}}\rangle$) 的最低能带上, 演化时间 T 后到达目标态 $|\psi_{\text{fin}}\rangle$[94].

4.3　高保真度复合绝热通道技术

本节主要介绍复合绝热通道 (composite adiabatic passage, CAP) 技术的应用, 主要包括两个方面: 一是将其应用到有限时间的两能级系统中, 以三角正、余弦函数形式的外场为例, 讨论了外场耦合强度、失谐强度等参数对跃迁概率的影响, 验证了该方法的有效性, 实现了有限时间两能级系统中高保真度布居数转移; 二是将

4.3 高保真度复合绝热通道技术

其推广到非线性两能级系统中, 考虑了非线性相互作用对该技术的影响, 发现只要控制脉冲足够多, 总能实现高保真度、快速以及具有参数鲁棒性的量子操控过程, 误差小于量子信息中的阈值 10^{-4}, 同时也指出了线性系统和非线性系统中该方法的异同.

利用外场实现量子态的操控在原子分子物理中至关重要, 可应用到测量学、干涉度量学、核磁共振、量子信息以及量子化学等领域 [2-4,59]. 特别在量子信息中, 要求这些操控技术具有如下三个重要特点: ① 高保真度, 容许误差要低于 10^{-4}, ② 快速性, 即为了避免退相干效应, 尽可能短地操控时间, ③ 鲁棒性, 即对系统或实验参数具有很好的鲁棒性 [57,64,95].

绝热通道技术 [adiabatic passage (AP) techniques] 是量子态操控中的一个重要工具. 近年来, 人们提出了许多绝热通道技术, 如快速绝热通道技术、精确绝热通道技术、受激拉曼绝热通道技术等 [1,96]. 该技术的优点是具有参数鲁棒性, 然而, 几乎所有的绝热通道技术的跃迁概率是不完全的. 量子操控的另一种鲁棒相干控制方法是复合脉冲技术, 该方法已经广泛应用于核磁共振 [97], 近几年也应用到量子光学以及量子信息过程 [98-100]. 这种方法用一系列具有特定相位的复合脉冲去代替传统的单个脉冲来实现两个量子态的跃迁, 这些相位作为控制工具在该技术中扮演着重要作用, 能使控制场为所需要的方式, 并且对实验参数, 如强度、频率等具有很好的鲁棒性. 该方法的缺点是脉冲面积不精确、频率补偿不好或者产生不需要的频率啁啾等 [101].

为了组合以上两种方法各自的优势, 实现鲁棒的、高保真度的量子控制, 人们相继提出了一系列优化方法 [8,11,13,14,30,102,103]. 复合绝热通道 (CAP) 技术就是其中一种重要而可行的方法 [8], 该方法用一系列具有特定相位的复合脉冲去代替单个脉冲, 通过控制脉冲的相位, 从而实现高保真度量子操控, 能实现完全的布居数转移, 远超过量子计算的容错标准. 近几年, 该方法引起了研究者的广泛关注 [104,105], 并且在稀土离子掺杂固体实验中得以验证 [106]. 然而, 一方面, 该方法主要用于无限时间量子系统的控制; 另一方面, 无论是理论工作, 还是实验验证, 主要局限于线性两能级系统, 粒子间相互作用并没有考虑. 在实际的量子操控中, 一方面, 人们总是希望控制时间是有限的, 如果一个脉冲持续时间很长, 则多个同样的脉冲去控制所需时间会很长; 另一方面, 粒子间相互作用, 特别是超冷原子系统中, 确实对跃迁动力学有极大的影响 [69,87]. 因此, 对有限的时间内完成的高保真度量子操控以及具有粒子间相互作用的非线性系统进行高保真度量子操控的研究具有重要的意义.

在本节中, 首先介绍复合绝热通道技术, 然后以三角正、余弦函数控制场为例, 将其推广到有限时间的两能级系统中 [107], 最后将该方法应用到具有粒子间相互作用的非线性两能级系统中 [15], 实现两能级系统的快速、高保真度、参数鲁棒性的量子操控.

4.3.1 复合绝热通道技术

考虑两能级系统的哈密顿量为

$$\boldsymbol{H}(t) = \frac{v(t)}{2}\hat{\sigma}_x + \frac{\gamma(t)}{2}\hat{\sigma}_z, \tag{4.37}$$

其中, $\hat{\sigma}_x$、$\hat{\sigma}_z$ 为泡利矩阵; $v(t)$、$\gamma(t)$ 分别表示施加外场的拉比频率 (耦合) 和失谐 (能级差) (在不同的物理系统中所代表物理意义不同). 该系统满足下述 Schrödinger 方程:

$$\mathrm{i}\frac{\partial}{\partial t}\boldsymbol{a}(t) = \boldsymbol{H}(t)\boldsymbol{a}(t), \tag{4.38}$$

其中, $\boldsymbol{a}(t) = [a_1(t), a_2(t)]^\mathrm{T}$ 为两态 $|1\rangle$ 和 $|2\rangle$ 的概率幅. 该模型尽管形式简单, 但是能描述许多物理现象. $\gamma(t)$ 和 $v(t)$ 选取不同的时间变化函数, 会对应一系列经典模型. 该系统的演化可用如下时间演化算符 $\boldsymbol{U}$ 来描述 [8]:

$$\boldsymbol{U} = \begin{pmatrix} a & b \\ -b^* & a^* \end{pmatrix}, \tag{4.39}$$

其中, a, b 是 Cayley-Klein 参数, 由初始时刻 t_i 的概率幅可以导出系统在任意时刻 t_f 的概率幅, 即 $\boldsymbol{a}(t_f) = \boldsymbol{U}(t_f, t_i)\boldsymbol{a}(t_i)$. 复合绝热通道技术要求对驱动场施加一个控制相位 ϕ, 演化算符 $\boldsymbol{U}$ 变为

$$\boldsymbol{U}_\phi = \begin{pmatrix} a & b\mathrm{e}^{-\mathrm{i}\phi} \\ -b^*\mathrm{e}^{\mathrm{i}\phi} & a^* \end{pmatrix}. \tag{4.40}$$

如果控制外场由 N 个完全一样的脉冲链组成, 每个脉冲具有一个控制相位 $\phi_n(n = 1, 2, \cdots, N)$, 则演化算符变为

$$\boldsymbol{U}_N = \boldsymbol{U}_{\phi_N}\boldsymbol{U}_{\phi_{N-1}}\cdots\boldsymbol{U}_{\phi_2}\boldsymbol{U}_{\phi_1}. \tag{4.41}$$

如果 $v(t)$ 是时间的偶函数, $\gamma(t)$ 是关于时间的奇函数, 通过适当选择控制相位 $\phi_k(k = 1, 2, \cdots, N)$, 就能实现高保真度量子操控, 如布居数反转. 一般脉冲数为奇数, 取 $N = 2n + 1$. 为简单起见, 文献 [8] 选取了一种具有对称性的控制相位, 并得到了其解析表达式:

$$\phi_k = \left(N + 1 - 2\left\lfloor\frac{n+1}{2}\right\rfloor\right)\left\lfloor\frac{n}{2}\right\rfloor\frac{\pi}{N}, \tag{4.42}$$

其中, 符号 $\lfloor x \rfloor$ 表示 floor 函数. 这里定义的跃迁概率为 $P = 1 - |a_1(t = t_f)|^2$, 即在初始时刻, 当初态在 $|1\rangle$ 态上时, 在 $t = t_f$ 时刻, 粒子最终能够跃迁到初始时处于空置状态 $|2\rangle$ 上的概率, 即为系统的保真度, $1 - P$ 为系统的误差.

4.3.2 有限时间两能级系统中高保真度布居数转移

本部分运用复合绝热通道技术, 研究了正、余弦函数形式外场驱动的两能级量子系统在有限时间内的跃迁问题, 实现了有限时间内该系统的高保真度布居数转移; 讨论了外场耦合强度、失谐强度等参数对跃迁概率的影响, 发现只要选择合适的控制相位, 在很大的参数范围内能够抑制跃迁概率的振荡, 保真度达到 1, 系统误差小于 10^{-4}, 从而实现高效、快速、稳定, 具有很好的参数鲁棒性的布居数完全转移; 接着运用经典哈密顿量分析和验证了该方法的有效性和可行性. 该技术适用于任何两能级系统, 也能推广到多能级系统中, 在量子信息、量子光学以及冷原子系统等领域有着广泛的应用[107].

两能级量子系统中布居数转移是原子、分子、光物理以及化学等领域的一个重要研究领域, 在高精密光谱[108,109]、分子动力学相干控制[110]、量子态制备、化学反应[2]、原子钟的设计[111], 特别在量子信息[5] 等领域有着极为广泛的应用前景. 长期以来, 在两能级系统中人们主要利用共振脉冲技术和各种绝热通道技术来实现两种态的布居数转移[52]. 共振脉冲技术保真度较高, 即转移效率高, 但是对实验及其环境参数敏感, 绝热通道技术尽管有很好的参数鲁棒性, 但往往保真度不高, 而且耗时长. 然而在量子计算中, 保真度的要求非常苛刻, 其容许误差应该控制在 10^{-4} 以内[5]. 因此, 探索既具有参数鲁棒性又具有高保真度的快速布居数转移技术成为多年来量子操控领域的一个热门研究课题. 如第 2 章所述, 近年来, 也涌现了一系列这方面的工作. 例如, Daems 等[95] 提出的单束整形脉冲 (single-shot shaped pulse) 技术; Demirplak 和 Rice[11,12], Berry [13] 等提出的超绝热技术; Chen 等提出的绝热捷径技术等[112,113]. 除此之外, Vitanov 研究组提出了一种复合绝热通道技术[8], 该技术拥有共振脉冲和绝热通道技术各自的优点, 既具有高保真度, 同时对实验参数有很好的鲁棒性, 成为量子操控中的一个重要手段. 该方法用一系列具有特定相位的复合脉冲代替单个脉冲, 通过控制这些脉冲的相位, 从而实现高保真度量子操控, 并且已经在实验上验证了这一方法的有效性[106].

两能级系统是研究量子现象的基本模型, 可用于研究一系列典型的量子问题[66,87,114–117]. 对两能级系统的布居数转移问题, 大部分工作主要集中在研究无限长时间 (infinite-time) 过程, 如 Landau-Zener 模型、Demkov-Kunike 模型、Nikiton 模型等[41], 对于有限时间的隧穿以及布居数转移问题研究相对较少[118,119]. 最近, Miao 等在氢原子的激光实验中应用了一种正、余弦函数形式的驱动外场, 研究了有限时间的两能级布居数转移问题[120], 引起了许多研究者的兴趣. 利用这种实验上易于实现的周期外场, Lu 等研究了一系列两能级系统布居数转移问题[121,122], 获得了一些解析结果. 然而, 其保真度不是很高, 且转移效率敏感地依赖于外场参数的变化. 本节以三角正、余弦函数形式的外场驱动两能级系统为例, 探讨了复合绝热通道技术在有限时间中的布居数转移问题,发现在有限的时间内,即使使用较少的脉冲个数,也能实现高保真度布居数转移,同时也研究了该方法在系统中的参数鲁棒性.

1. 模型

文献 [8] 和 [123] 等已经研究了在无限长时间外场中该方法的有效性. 然而, 在量子操控中人们总是希望控制时间是有限的, 这种有限的时间内完成的高保真度量子操控在量子计算中具有重要的意义. 本节选择文献 [120]～[124] 所采用三角函数形式驱动场:

$$v(t) = v_0 \cos(\omega t), \quad \gamma(t) = \gamma_0 \sin(\omega t), \tag{4.43}$$

即耦合是余弦函数形式的脉冲 (初始时刻和结束时刻强度均为零), 失谐是时间的正弦函数, 其中 $\omega = \pi/T$, T 是外场周期, v_0 和 γ_0 分别是耦合强度和失谐强度, 系统演化要求在有限时间 $[-T/2, T/2]$ 进行. 该外场是一种典型的有限时间模型, 易于实验操控, 当外场周期 T 很大时, 退化为经典的 Landau-Zener 模型. 它对应的量子系统的本征能级为 $E_\pm = \pm \frac{1}{2} \sqrt{v_0^2 \cos^2(\omega t) + \gamma_0^2 \sin^2(\omega t)}$, 不同参数下的外场和对应的能级如图 4.28 所示.

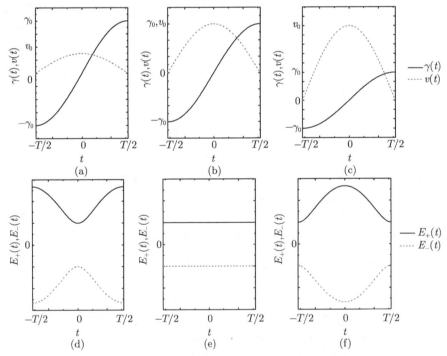

图 4.28 三角正、余弦函数形式的外场和对应的能级结构

(a), (d) $\gamma_0 > v_0$; (b), (e) $\gamma_0 = v_0$; (c), (f) $\gamma_0 < v_0$

4.3 高保真度复合绝热通道技术

2. 高保真度布居数转移

本部分分别计算了不同情况下系统的跃迁概率 P 随参数的变化情况,计算中外场参数 v_0 和 γ_0 均用 ω 去无量纲化,每个脉冲的持续时间为 $[-T/2, T/2]$,所有的计算中取 $T = \pi$.

图 4.29 显示了在不同脉冲个数下,跃迁概率 P 随参数 v_0 和 γ_0 的变化情况. 当脉冲个数 $N = 1$ 时,就是文献 [120]~ [122] 研究过的情况,即没有控制相位的正、余弦函数外场的两能级隧穿问题,跃迁概率对应的区域分为两部分:一部分为振荡区域;另一部分为非振荡区域. 可以看到,大部分区域跃迁概率 P 是振荡的,$P > 0.9999$ 的区域只是其中非常小的一小部分,如图 4.29(a) 所示. 这种情况下要实现布居数绝热转移,首先演化时间会非常长,其次还要满足绝热条件. 增加脉冲个数,当 $N = 3$ 时,对应的控制相位分别为 $\phi_1 = \phi_3 = 0, \phi_2 = 2\pi/3$,振荡区域依然仅存在于 γ_0 和 v_0 取值比较小的区域,$P > 0.9999$ 的区域已经扩大. 继续增加脉冲个数,当 $N = 5$ 时,对应的控制相位分别为 $\phi_1 = \phi_5 = 0, \phi_2 = \phi_4 = 4\pi/5$, $\phi_3 = 2\pi/5$,振荡区域继续减小,$P > 0.9999$ 的区域进一步增大,几乎布满了大部分 γ_0 和 v_0 的参数区域. 这种通过调节相位的多脉冲操控方式,不仅抑制了跃迁概率的振荡,而且也提高了跃迁概率. 说明利用这种方法确实能实现高保真度跃迁,完成布居数完全转移,且随着脉冲数量的增加,高跃迁概率的参数范围会进一步扩大.

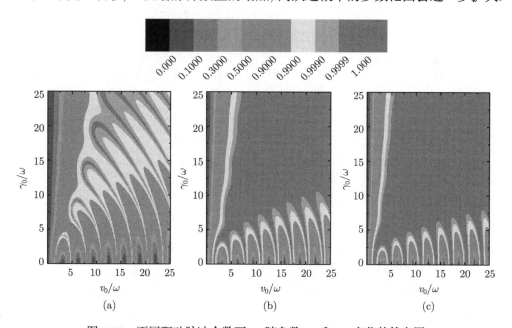

图 4.29　不同驱动脉冲个数下 P 随参数 v_0 和 γ_0 变化的等高图
(a) $N = 1$; (b) $N = 3$; (c) $N = 5$

为了进一步讨论这种操控方式对外场参数的鲁棒性, 也分别计算了 P 和 $1-P$ 分别随参数 v_0 和 γ_0 的变化情况, 如图 4.30 所示, 图 4.30(a) 和 (b) 中 $\gamma_0/\omega = 8$, (c) 和 (d) 中 $v_0/\omega = 8$. 可以发现, 在脉冲个数为 1 时, 保真度随 v_0 和 γ_0 均是振荡的, 随着脉冲数增加, 通过相位的调节, 这些振荡逐渐减弱, 甚至完全消失, 即使使用比较少的脉冲个数, 如 3 个或者 5 个的情况下, 复合绝热通道技术也能实现高保真度布居数转移, 其系统误差在很大的参数区域都在 10^{-4} 以下, 特别是 $N=5$ 时, 在计算的参数区域内, 系统误差全部在 10^{-4} 以下. 这就充分说明复合绝热通道技术具有很好的参数鲁棒性, 也进一步验证了以上的结论.

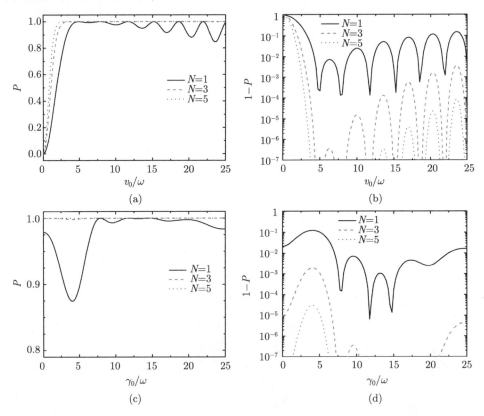

图 4.30 不同脉冲个数下 P 和 $1-P$ 分别随参数 v_0 和 γ_0 的变化情况

(a) $P-v_0$; (b) $(1-P)-v_0$; (c) $P-\gamma_0$; (d) $(1-P)-\gamma_0$

为了更加清晰地认识这种操控技术的优越性, 探索操控过程中的物理机制, 引入系统的经典哈密顿量[87]. 首先定义一对正则变量: 跃迁概率 P 和相对相位 θ, 其中 $P = |a_2|^2, a_1 = \sqrt{1-P} \exp(\mathrm{i}\theta_1), a_2 = \sqrt{P} \exp(\mathrm{i}\theta_2), \theta = \theta_2 - \theta_1$, 则系统 (4.38) 对应的经典哈密顿量如下:

4.3 高保真度复合绝热通道技术

$$H(t) = \frac{\gamma(t)}{2}(1 - 2P) + v(t)\sqrt{P(1-P)}\cos(\theta + \phi_k), \qquad (4.44)$$

$\dot{\theta} = \partial H/\partial P, \dot{P} = -\partial H/\partial \theta$,可以得到一对正则方程:

$$\dot{P} = v(t)\sqrt{P(1-P)}\sin(\theta + \phi_k), \qquad (4.45)$$

$$\dot{\theta} = -\gamma(t) + \frac{v(t)(1-2P)}{2\sqrt{P(1-P)}}\cos(\theta + \phi_k). \qquad (4.46)$$

借助上述正则方程,计算了取 5 个脉冲时的跃迁概率 P,相对相位 θ,能量 H 以及控制相位 ϕ_k 随时间的演化情况,如图 4.31 所示. 计算时取 $\gamma_0/\omega = 10, v_0/\omega = 3$, $t_i (i = 1, 2, \cdots, 5)$ 为每个脉冲的中心位置对应的时间. 只有一个脉冲时,跃迁概率还比较低,三个脉冲结束时,跃迁概率已经有了大幅度提高,随着脉冲数量的增多,最后的跃迁概率会变大,达到要求的精度,操控过程中每个脉冲的控制相位发挥了重要作用. 为了比较,也计算了没有控制相位时, P、θ、H 随时间的演化情况. 同时也通过直接解 Schrödinger 方程 (4.38) 计算了 P 随时间的演化情况 [如图 4.31(a) 中圆圈线],二者符合得很好. 值得注意的是,控制相位的出现,会调控系统演化过程中的经典能量,使得控制开始时刻和结束时刻对应的能量相同,这也满足绝热演

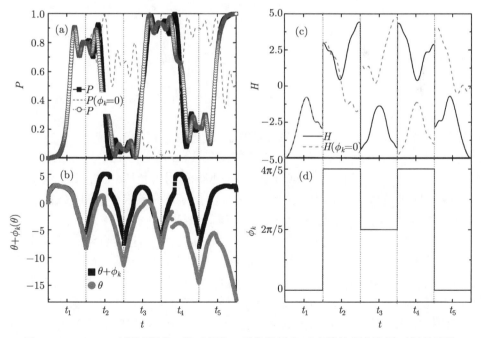

图 4.31 $N = 5$ 时跃迁概率、相对相位、系统能量和对应的控制相位随时间的演化
(a) P-t; (b) θ-t; (c) H-t; (d) ϕ_k-t

化的基本要求[125]. 操控结束时, 系统能量为 $H = -\gamma/2$, 使得 $P \approx 1$, 又由于此时 $|a_1|^2$ 已经很小, 演化算符矩阵元 $U_{11}^{(N)} = U_{11}^{(5)} = a^5$, 这样 $P = 1 - |a_1|^{2N} = 1 - |a_1|^{10}$, 其值几乎为 $1^{[8]}$, 从而实现高保真度布居数转移. 可以看到, 随着时间的变化, 跃迁概率也会出现振荡, 发生干涉现象, 这和外场驱动形式有关[41], 但是在脉冲结束的最后时刻, 跃迁概率都很高, 有极高的保真度[107].

4.3.3 非线性两能级系统中的应用

本节将复合绝热通道技术应用于非线性两能级系统, 研究发现, 粒子间相互作用倾向于增加脉冲个数, 即只要有足够多的脉冲数, 高保真度跃迁过程总能被实现. 对于一般的绝热过程, 由于要实现绝热条件, 一个脉冲持续时间也很长, 对于线性系统中的复合绝热通道技术, 总持续时间是单个脉冲的 N 倍, 所需时间原则上是无穷长, 这在物理上是不实际的. 因而, 不同于线性系统, 在非线性系统中, 为了实现快速的量子操控过程, 总保持脉冲持续时间固定[15].

1. 模型

本节考虑的非线性两态量子系统由如下无量纲化的 Schrödinger 方程描述[72]:

$$\mathrm{i}\frac{\partial}{\partial t}\boldsymbol{a}(t) = \boldsymbol{H}(t)\boldsymbol{a}(t), \tag{4.47}$$

其中哈密顿量为

$$\boldsymbol{H}(t) = \frac{v(t)}{2}\hat{\sigma}_x + \left[\frac{\gamma(t)}{2} + \frac{c}{2}(|a_2|^2 - |a_1|^2)\right]\hat{\sigma}_z, \tag{4.48}$$

同样, 其中, $\boldsymbol{a}(t) = [a_1(t), a_2(t)]^\mathrm{T}$ 为两态概率幅; $\hat{\sigma}_x$ 和 $\hat{\sigma}_z$ 为泡利算符; $\gamma(t)$ 和 $v(t)$ 分别是两态能级差和耦合强度; c 代表非线性参数, 用来描述粒子间相互作用. 总概率守恒, 即 $|a_1|^2 + |a_2|^2 = 1$. 该模型不仅在理论上具有重要意义, 更为重要的是能广泛应用在许多物理系统中, 如用于描述纳米磁子的自旋隧穿[126]、双势阱或者光晶格中的玻色-爱因斯坦凝聚[127,128] 以及耦合波导管[129] 等.

由 4.3.2 节讨论知道, 在线性情况下, 即 $c = 0$ 时, 人们对复合绝热通道技术已经做了深入的研究[8], 并展现了其非凡的能力[105,106]. 该方法中两态系统的演化算符能够由 Cayley-Klein 参数表示, 并且驱动场中由适当相位的一系列脉冲代替单个脉冲. 这样能够抑制跃迁过程中的非绝热振荡, 甚至在 3~5 个复合脉冲的情况下就能使控制误差低于量子计算的误差阈值 10^{-4}. 另外, 只要控制场满足上述对称性关系, 其控制相位不依赖于脉冲形状和啁啾形式, 那么在上述非线性系统中, 这种方法是否适用呢?

2. 非线性效应

本部分考虑非线性系统中的复合绝热通道技术, 研究粒子间相互作用对复合绝热通道技术的影响. 由于系统中出现了非线性, Schrödinger 方程 (4.47) 不再能够解析求解, 因此采用 4-5 阶 Runge-Kutta 法进行数值求解, 研究其隧穿动力学. 为了实现高保真度的量子跃迁, 采用具有 N ($N = 2n + 1$, n 是整数) 个脉冲的脉冲链, 每一个脉冲具有一个特定相位 $\phi_k (k = 1, 2, \cdots, N)$, 并且加载在耦合强度上, 即 $v(t) \to v(t) e^{i\phi_k}$. 为了简单起见, 在非线性系统中, 控制相位仍然取线性系统中的形式 [8].

$$\phi_k = \left(N + 1 - 2\left\lfloor \frac{k+1}{2} \right\rfloor\right) \left\lfloor \frac{k}{2} \right\rfloor \frac{\pi}{N}, \tag{4.49}$$

同样 $\lfloor x \rfloor$ 表示 floor 函数, 并且相位具有对称性, 即 $\phi_k = \phi_{N+1-k}$, $\phi_1 = \phi_N = 0$, 并且要求耦合强度 $v(t)$ 是时间的偶函数, 能级差或者失谐 $\gamma(t)$ 是时间的奇函数.

作为一个例子, 考虑了 Allen-Eberly (AE) 模型, 其中耦合强度为 sech 形式, 而能级差为 tanh 形式 [8],

$$v(t) = v_0 \text{sech}(t/T), \quad \gamma(t) = \alpha \tanh(t/T), \tag{4.50}$$

其中, v_0 和 α 为常参数 (具有频率的量纲); T 是脉冲宽度.

在线性情况下, 该模型的跃迁概率 $P = |a_2|^2 = 1 - |a|^2$ 为 [8]

$$P = \frac{\cosh(\pi\alpha T) - \cos\left(\pi T \sqrt{v_0^2 - \alpha^2}\right)}{1 + \cosh(\pi v_0 T)}$$

$$= 1 - \frac{\cos^2\left(\frac{1}{2}\pi T \sqrt{v_0^2 - \alpha^2}\right)}{\cosh^2\left(\frac{\pi\alpha T}{2}\right)}. \tag{4.51}$$

当 $v_0 < \alpha$ 时, 方程 (4.51) 中的余弦函数用双曲余弦函数代替. 要实现跃迁概率 $P = 1$ (即完全的布居数转移), 必须满足 $\sqrt{v_0^2 - \alpha^2} T = 2n + 1$, 这里 $n = 0, 1, 2, \cdots$ (整数). 绝热极限下 ($v_0 > \alpha \gg 2/T$), 跃迁概率也倾向于 1. 如果 α 不是足够大, 在 v_0 方向上将出现非绝热振荡, 因而使跃迁概率降低. 通过运用复合绝热通道技术, 即使在 3~5 个脉冲的情况下, 也能大幅度抵消非绝热振荡, 从而实现高保真度的量子跃迁. 所有变量用 T 重新无量纲化, 计算时取 $T = 1$、v_0, α 和 c 的单位均是 $1/T$.

图 4.32 显示 AE 模型中跃迁概率随复合脉冲个数 N 和粒子间相互作用 c 的变化情况, 其中 $\alpha = 1, v_0 = 1.2$. 显然, 粒子间相互作用会强烈地影响系统的跃迁动力学. 对于弱的相互作用, 即使在 3~5 个复合脉冲的情况下, 高的跃迁概率也能够

实现. 随着非线性相互作用的增加, 非绝热振荡也逐渐加强, 从而使得跃迁概率大大降低, 这就意味着在少的控制数量下, 高跃迁概率不再实现. 然而, 有意思的是, 只要复合脉冲数量足够多, 这些非绝热振荡就能够被抑制, 也就是说, 只要控制脉冲足够多, 复合绝热通道技术也能够应用于非线性两能级系统. 值得注意的是, 线性系统中, 控制脉冲总的持续时间是单个脉冲时间的 N 倍. 然而, 在非线性系统中, 保持总的控制脉冲持续时间不变, 即不论脉冲数量是多少, 总是固定总的持续时间. 在所有的数值模拟中, 取时间为 $-100 \sim 100$.

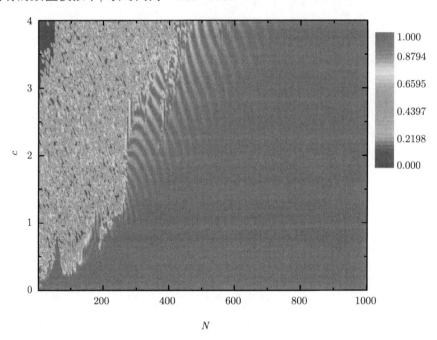

图 4.32 AE 模型中跃迁概率随复合脉冲个数 N 和粒子间相互作用 c 的变化情况

图 4.33 给出了在不同复合脉冲个数 N 时、跃迁概率 P 随拉比频率峰值 v_0 变化情况. 其中图 4.33(a) 和 (b) 分别显示相互作用 $c = 0.2$ 时 1299 个控制脉冲对应的跃迁概率和误差情况, 而图 4.33(c) 和 (d) 表示相互作用 $c = 2.0$ 时对应的情况, 图中 $\alpha = 1$. 可以发现非绝热振荡大大被抑制, 其误差低于量子信息中的阈值 10^{-4}. 为了和文献 [8] 中线性情况作对比, 也画出了 $c = 0$, $N = 5$ 时跃迁概率随 v_0 的变化 (虚线). 在线性情况下, 只要 5 个控制脉冲就足以实现极高的保真度. 然而, 在非线性系统中, 高保真度尽管可以实现, 并且其误差低于 10^{-4}, 但是需要更多的控制脉冲.

4.3 高保真度复合绝热通道技术

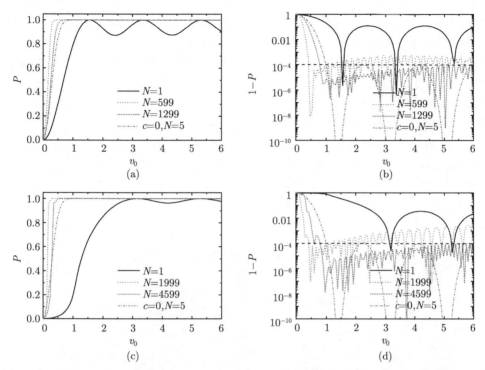

图 4.33 不同相互作用 $c = 0.2$ (a), $c = 2.0$ (c) 和复合脉冲个数 N 时, 跃迁概率随拉比峰值 v_0 变化情况, (b), (d) 表示对应的误差变化情况

为了检验该方法对场参数变量的鲁棒性, 计算了不同参数下跃迁概率随啁啾率 α 和拉比频率峰值 v_0 的变化情况, 如图 4.34 所示. 可以看出, 复合绝热通道技术具有很好的参数鲁棒性, 并且相比于单个脉冲的情况, 多脉冲使得高保真度参数范围大大扩展. 不同于线性系统, 在非线性系统中, 为了实现高保真度, 确实需要多个复合控制脉冲, 并且随着非线性相互作用的增强, 要达到同样的结果, 需要的脉冲数量会逐渐增多.

为进一步探索上述现象, 引入了一对正则变量: 相对相位 $\theta = \theta_2 - \theta_1$ 和跃迁概率 $P = |a_2|^2$, 这里 $a_1 = \sqrt{1-P}\exp(\mathrm{i}\theta_1)$, $a_2 = \sqrt{P}\exp(\mathrm{i}\theta_2)$. 这样可以由正则方程 $\mathrm{d}P/\mathrm{d}t = -\partial H/\partial \theta$, $\mathrm{d}\theta/\mathrm{d}t = \partial H/\partial P$ 获得等效经典哈密顿量,

$$H(t) = \frac{\gamma}{2}(1-2P) - \frac{c}{4}(1-2P)^2 + v\sqrt{P(1-P)}\cos(\theta + \phi_k). \quad (4.52)$$

这个经典哈密顿量能完全描述系统 (4.47) 的动力学性质[87]. 图 4.35 显示了不同的相互作用和脉冲数量情况下, AE 模型中跃迁概率和相对相位分别在具有控制相位和没有控制相位时的演化情况. 在图 4.35(a)~(c) 中, 给出了相空间轨迹, 黑色三角和星分别代表初态和终态. 可以看到, 控制相位在复合脉冲技术中起到了关键

作用, 它能极大地影响系统的跃迁动力学. 通过选择合适的控制相位, 就可以抑制跃迁概率的非绝热振荡, 从而实现复合绝热通道过程中的高保真度. 在线性情况下, 复合绝热通道技术在少脉冲个数时就能达到很好的操控效果, 并且每一个脉冲都能使布居数发生大的改变, 但是不能实现完全转移, 直到最后干涉相消效应使得系统实现完全的布居数转移. 而在非线性系统中, 复合绝热通道技术需要很多的复合脉冲, 每一个都会使布居数有一些小的改变, 最终也能实现高保真度. 这里采用的控制相位是相同的. 通过直接解 Schrödinger 方程 (4.47), 也能重复上面的结果. 相比于文献 [8] 中的方法, 本节的方法中, 总的脉冲持续时间是保持不变的, 这样不同的脉冲个数控制脉冲的形式也不同. 例如, 对于 1299 脉冲的 AE 模型, 实际上就变成了有限持续时间的 Landau-Zener 模型 [图 4.35(d)], 如果没有控制相位, 跃迁概率就会出现振荡.

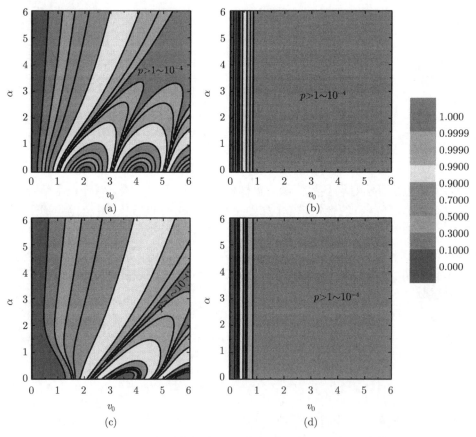

图 4.34 跃迁概率随啁啾率 α 和拉比频率峰值 v_0 的变化情况

(a)、(b) $c = 0.2$, (c)、(d) $c = 2.0$

(a) $N = 1$; (b) $N = 1399$; (c) $N = 1$; (d) $N = 4799$

4.3　高保真度复合绝热通道技术

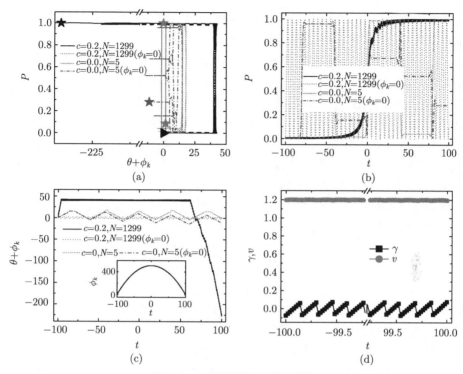

图 4.35　相图及其跃迁动力学

(a) 相图; (b) 跃迁概率的时间演化 (子图: 控制相位的时间演化); (c) 相对相位的时间演化; (d) 1299 脉冲序列时 γ, v 随时间的演化

4.3.4　结论与讨论

复合绝热通道技术应用于两个方面.

一是应用在三角函数形式外场驱动的两能级系统中, 发现在有限的时间内能实现该系统的布居数完全转移, 而且保真度高, 系统误差小于 10^{-4}. 通过讨论各个外场参数对跃迁概率的影响, 发现只要控制相位合理, 就能在大部分参数范围内实现这种转移, 该技术具有很好的参数鲁棒性. 最后利用经典哈密顿量, 分析了跃迁概率、相对相位、系统能量以及控制相位随时间的演化, 进一步分析了这种技术的优越性及其物理机制. 这种方法和目前其他的优化控制方法既有相似之处, 即都能实现高保真度、具有很好的参数鲁棒性和快速性的量子操控, 然而又完全不同. 例如, 单束整形脉冲技术, 也需要通过控制相位来进行量子操控, 其相位可根据不同鲁棒性展开成傅里叶级数的形式, 通过确定展开系数来实现操控. 该方法只需要一束脉冲即可, 只是不同鲁棒性要求其脉冲形式略有不同, 而复合绝热通道技术所采用的是形状完全相同, 但具有不同相位的多个脉冲. 超绝热技术是通过附加哈密顿量以

抵消非绝热振荡,绝热捷径技术则是基于不变量的反控制方法等来实现高保真度绝热量子操控. 这种在有限的时间内实现高保真度的布居数转移技术,适用于任何两能级系统,也能推广到多能级系统中,相信在量子计算、量子光学、冷原子系统等领域有着广泛的应用 [107].

二是应用在非线性两能级系统中,分析了粒子间相互作用对复合绝热通道技术的影响. 发现非线性相互作用倾向于增加脉冲个数,只要脉冲数量足够多,该技术同样可以抑制非绝热振荡,使得误差低于量子信息中的阈值 10^{-4}. 和线性系统不同的是,不论脉冲个数是多少,都固定了总脉冲持续时间,从而能实现快速高保真度量子操控. 这些性质使得复合绝热通道技术成为超高保真度、超快量子控制中的一种重要工具. 这种技术也可以在加速光晶格玻色-爱因斯坦凝聚体实验中实现,其中,粒子间相互作用也可以通过 Feshbach 共振任意调节,而且凝聚体高密度也能实现,以至于非线性效应在这些系统中能够被观察到 [31,66,130]. 近年来,周期势中玻色-爱因斯坦凝聚两能带之间的非线性 Landau-Zener 隧穿已经在实验中被观察到 [71,74],加之超短激光脉冲也已经应用于量子控制中 [1,52],因此,可以相信高保真度复合绝热通道技术也能够在实验上实现. 需要指出的是,在非线性系统中,仍然采用线性系统中的控制相位,这也许不是最优化的形式. 然而,在非线性系统的高保真度量子操控中,这种选择却是最简单,也是最方便的. 如何选择非线性系统中更加合适的控制相位仍然是值得研究的一个极具挑战的开放问题.

参 考 文 献

[1] KRÁl P, THANOPULOS I, SHAPIRO M. Colloquium: Coherently controlled adiabatic passage[J]. Rev. Mod. Phys., 2007, 79(1): 53.

[2] BERGMANN K, THEUER H, SHORE B W. Coherent population transfer among quantum states of atoms and molecules[J]. Rev. Mod. Phys., 1998, 70: 1003-1025.

[3] BRIF C, CHAKRABARTI R, RABITZ H. Control of quantum phenomena: Past, present and future[J]. New J. Phys., 2010, 12(7): 075008.

[4] SABBAH H, BIENNIER L, SIMS I R, et al. Understanding reactivity at very low temperatures: The reactions of oxygen atoms with alkenes[J]. Science, 2007, 317(5834): 102-105.

[5] NIELEN M, CHUANG I. Quantum Computation and Quantum Information[M]. Cambridge: Cambridge Univ. Press, 2000.

[6] REZAKHANI A, KUO W J, HAMMA A, et al. Quantum adiabatic brachistochrone[J]. Phys. Rev. Lett., 2009, 103(8): 080502.

[7] DU J, HU L, WANG Y, et al. Experimental study of the validity of quantitative conditions in the quantum adiabatic theorem[J]. Phys. Rev. Lett., 2008, 101(6): 060403.

[8] TOROSOV B T, GUÉRIN S, VITANOV N V. High-fidelity adiabatic passage by composite sequences of chirped pulses[J]. Phys. Rev. Lett., 2011, 106: 233001.

[9] EMMANOUILIDOU A, ZHAO X G, AO P, et al. Steering an eigenstate to a destination[J]. Phys. Rev. Lett., 2000, 85: 1626-1629.

[10] MASUDA S, NAKAMURA K. Acceleration of adiabatic quantum dynamics in electromagnetic fields[J]. Phys. Rev. A, 2011, 84(4): 043434.

[11] DEMIRPLAK M, RICE S A. Adiabatic population transfer with control fields[J]. The Journal of Physical Chemistry A, 2003, 107(46): 9937-9945.

[12] DEMIRPLAK M, RICE S A. On the consistency, extremal, and global properties of counterdiabatic fields[J]. J. Chem. Phys., 2008, 129(15): 154111.

[13] BERRY M. Transitionless quantum driving[J]. J.Phys. A: Mathematical and Theoretical, 2009, 42(36): 365303.

[14] TORRONTEGUI E, IBÁNEZ S, MARTÍNEZ-Garaot S, et al. Shortcuts to adiabaticity[J]. Adv. At. Mol. Opt. Phys, 2013, 62:117-169.

[15] DOU F Q, CAO H, LIU J, et al. High-fidelity composite adiabatic passage in nonlinear two-level systems[J]. Phys. Rev. A, 2016, 93(4): 043419.

[16] DEMIRPLAK M, RICE S A. Assisted adiabatic passage revisited[J]. J. Phys. Chem. B, 2005, 109(14): 6838-6844.

[17] LIM R, BERRY M. Superadiabatic tracking of quantum evolution[J]. J. Phys. A: Mathematical and General, 1991, 24(14): 3255.

[18] CHEN X, RUSCHHAUPT A, SCHMIDT S, et al. Fast optimal frictionless atom cooling in harmonic traps: Shortcut to adiabaticity[J]. Phys. Rev. Lett., 2010, 104(6): 063002.

[19] DEL CAMPO A. Shortcuts to adiabaticity by counterdiabatic driving[J]. Phys. Rev. Lett., 2013, 111(10): 100502.

[20] LU M, XIA Y, SHEN L T, et al. Shortcuts to adiabatic passage for population transfer and maximum entanglement creation between two atoms in a cavity[J]. Phys. Rev. A, 2014, 89(1): 012326.

[21] TAKAHASHI K. Transitionless quantum driving for spin systems[J]. Phys. Rev. E, 2013, 87(6): 062117.

[22] PAUL K, SARMA A K. Shortcut to adiabatic passage in a waveguide coupler with a complex-hyperbolic-secant scheme[J]. Phys. Rev. A, 2015, 91(5): 053406.

[23] MASUDA S, GÜNGÖRDÜ U, CHEN X, et al. Fast control of topological vortex formation in Bose-Einstein condensates by counterdiabatic driving[J]. Phys. Rev. A, 2016, 93(1): 013626.

[24] SANTOS A C, SILVA R D, SARANDY M S. Shortcut to adiabatic gate teleportation[J]. Phys. Rev. A, 2016, 93(1): 012311.

[25] SUN Z, ZHOU L, XIAO G, et al. Finite-time Landau-Zener processes and counterdiabatic driving in open systems: Beyond Born, Markov, and rotating-wave approximations[J]. Phys. Rev. A, 2016, 93(1): 012121.

[26] LIANG Z T, YUE X, LV Q, et al. Proposal for implementing universal superadiabatic geometric quantum gates in nitrogen-vacancy centers[J]. Phys. Rev. A, 2016, 93(4): 040305.

[27] OKUYAMA M, TAKAHASHI K. From classical nonlinear integrable systems to quantum

shortcuts to adiabaticity[J]. Phys. Rev. Lett., 2016, 117(7): 070401.

[28] IBÁÑEZ S, MARTÍNEZ-GARAOT S, CHEN X, et al. Shortcuts to adiabaticity for non-Hermitian systems[J]. Phys. Rev. A, 2011, 84(2): 023415.

[29] SCHAFF J F, SONG X L, VIGNOLO P, et al. Fast optimal transition between two equilibrium states[J]. Phys. Rev. A, 2010, 82(3): 033430.

[30] BASON M G, VITEAU M, MALOSSI N, et al. High-fidelity quantum driving[J]. Nature Phys., 2012, 8: 147-152.

[31] MALOSSI N, BASON M G, VITEAU M, et al. Quantum driving protocols for a two-level system: From generalized Landau-Zener sweeps to transitionless control[J]. Phys. Rev. A, 2013, 87: 012116.

[32] ZHANG J, SHIM J H, NIEMEYER I, et al. Experimental implementation of assisted quantum adiabatic passage in a single spin[J]. Phys. Rev. Lett., 2013, 110(24): 240501.

[33] DU Y X, LIANG Z T, LI Y C, et al. Experimental realization of stimulated Raman shortcut-to-adiabatic passage with cold atoms[J]. Nature commun., 2016, 7: 12479.

[34] AN S, LV D, DEL CAMPO A, et al. Shortcuts to adiabaticity by counterdiabatic driving for trapped-ion displacement in phase space[J]. Nature commun., 2016, 7: 12999.

[35] ZHOU B B, BAKSIC A, RIBEIRO H, et al. Accelerated quantum control using superadiabatic dynamics in a solid-state lambda system[J]. Nature Phys., 2017, 13(4): 330-334.

[36] LANDAU L D. On the theory of transfer of energy at collisions II[J]. Phys. Z. Sowjetunion, 1932, 2: 46-51.

[37] ZENER C. Non-adiabatic crossing of energy levels[J]. Proc. R. Soc. London, 1932, 137: 696-702.

[38] STÜCKELBERG E C G. Theory of inelastic collisions between atoms (Theory of inelastic collisions between atoms, using two simultaneous differential equations)[J]. Helv. Phys. Acta, 1932, 5: 369-422.

[39] MAJORANA E. Atomi orientati in campo magnetico variabile[J]. Nuovo Cimento, 1932, 9: 43-50.

[40] ROLAND J, CERF N J. Quantum search by local adiabatic evolution[J]. Phys. Rev. A, 2002, 65(4): 042308.

[41] DOU F Q, LI S C, CAO H. Combined effects of particle interaction and nonlinear sweep on Landau–Zener transition[J]. Phys. Lett. A, 2011, 376(1): 51-55.

[42] GARANIN D A, SCHILLING R. Effects of nonlinear sweep in the Landau-Zener-Stueckelberg effect[J]. Phys. Rev. B, 2002, 66: 174438.

[43] DOU F Q, LIU J, FU L B. High-fidelity superadiabatic population transfer of a two-level system with a linearly chirped Gaussian pulse[J]. EPL (Europhysics Lett.), 2017, 116(6): 60014.

[44] 冯平, 孙建安, 王文元, 等. Demkov-Kunike 模型的高保真度超绝热量子驱动 [J]. 激光与光电子学进展, 2017, 54(12): 122701.

[45] MASUDA S, RICE S A. Selective vibrational population transfer using combined stimulated

Raman adiabatic passage and counter-diabatic fields[J]. J. Phys. Chem. C, 2014, 119 (26): 14513-14523.

[46] OSVAY K, ROSS I N. Efficient tuneable bandwidth frequency mixing using chirped pulses[J]. Opt. Commun., 1999, 166(1): 113-119.

[47] ZHAO X M, DIELS J C, WANG C Y, et al. Femtosecond ultraviolet laser pulse induced lightning discharges in gases[J]. IEEE Journal of Quantum Electronics, 1995, 31(3): 599-612.

[48] KHACHATRYAN A, VAN GOOR F, VERSCHUUR J W, et al. Effect of frequency variation on electromagnetic pulse interaction with charges and plasma[J]. Phys. Plasm., 2005, 12(6): 062116.

[49] HAUSE A, HARTWIG H, SEIFERT B, et al. Phase structure of soliton molecules[J]. Phys. Rev. A, 2007, 75(6): 063836.

[50] SCHMIDGALL E, EASTHAM P, PHILLIPS R. Population inversion in quantum dot ensembles via adiabatic rapid passage[J]. Phys. Rev. B, 2010, 81(19): 195306.

[51] VASILEV G, VITANOV N. Coherent excitation of a two-state system by a linearly chirped Gaussian pulse[J]. J. Chem. Phys., 2005, 123: 174106.

[52] GUÉRIN S, HAKOBYAN V, JAUSLIN H. Optimal adiabatic passage by shaped pulses: Efficiency and robustness[J]. Phys. Rev. A, 2011, 84(1): 013423.

[53] MALINOVSKY V, KRAUSE J. General theory of population transfer by adiabatic rapid passage with intense, chirped laser pulses[J]. Eur. Phys. J. D-Atomic, Molecular, Optical and Plasma Physics, 2001, 14(2): 147-155.

[54] DEMKOV Y N, KUNIKE M. Hypergeometric models for the two-state approximation in collision theory[J]. Vestn. Leningr. Univ., 1969, 16 (3): 39.

[55] SUOMINEN K A, GARRAWAY B. Population transfer in a level-crossing model with two time scales[J]. Phys. Rev. A, 1992, 45(1): 374.

[56] LACOUR X, GUERIN S, YATSENKO L, et al. Uniform analytic description of dephasing effects in two-state transitions[J]. Phys. Rev. A, 2007, 75(3): 033417.

[57] HOLLENBERG L C. Quantum control: through the quantum chicane[J]. Nature Phys., 2012, 8(2): 113-114.

[58] LADD T D, JELEZKO F, LAFLAMME R, et al. Quantum computers[J]. Nature, 2010, 464(7285): 45-53.

[59] RICE S A, ZHAO M. Optical Control of Molecular Dynamics[M]. New York: John Wiley, 2000.

[60] BARTELS R A, WEINACHT T C, WAGNER N, et al. Phase modulation of ultrashort light pulses using molecular rotational wave packets[J]. Phys. Rev. Lett., 2001, 88: 013903.

[61] DE MELO C A S. When fermions become bosons: Pairing in ultracold gases[J]. Physics Today, 2008, 61: 45.

[62] FLAMBAUM V V, KOZLOV M G. Enhanced sensitivity to the time variation of the fine-structure constant and m_p/m_e in diatomic molecules[J]. Phys. Rev. Lett., 2007, 99: 150801.

[63] HÄNSCH T W. Nobel lecture: Passion for precision[J]. Rev. Mod. Phys., 2006, 78: 1297-1309.

[64] SCHLOSSHAUER M. Decoherence, the measurement problem, and interpretations of quantum mechanics[J]. Rev. Mod. Phys., 2005, 76: 1267-1305.

[65] CANEVA T, MURPHY M, CALARCO T, et al. Optimal control at the quantum speed limit[J]. Phys. Rev. Lett., 2009, 103: 240501.

[66] DOU F, FU L, LIU J. High-fidelity fast quantum driving in nonlinear systems[J]. Phys. Rev. A, 2014, 89(1): 012123.

[67] FU L, LIU J. Quantum entanglement manifestation of transition to nonlinear self-trapping for Bose-Einstein condensates in a symmetric double well[J]. Phys. Rev. A, 2006, 74: 063614.

[68] LIU J, FU L, OU B Y, et al. Theory of nonlinear Landau-Zener tunneling[J]. Phys. Rev. A, 2002, 66(2): 023404.

[69] TRIMBORN F, WITTHAUT D, KEGEL V, et al. Nonlinear Landau–Zener tunneling in quantum phase space[J]. N. J. Phys., 2010, 12(5): 053010.

[70] ZOBAY O, GARRAWAY B M. Time-dependent tunneling of Bose-Einstein condensates[J]. Phys. Rev. A, 2000, 61: 033603.

[71] CHEN Y A, HUBER S D, TROTZKY S, et al. Many-body Landau-Zener dynamics in coupled one-dimensional Bose liquids[J]. Nature Phys., 2010, 7(1): 61-67.

[72] WU B, NIU Q. Nonlinear Landau-Zener tunneling[J]. Phys. Rev. A, 2000, 61: 023402.

[73] CRISTIANI M, MORSCH O, MÜLLER J H, et al. Experimental properties of Bose-Einstein condensates in one-dimensional optical lattices: Bloch oscillations, Landau-Zener tunneling, and mean-field effects[J]. Phys. Rev. A, 2002, 65: 063612.

[74] JONA-LASINIO M, MORSCH O, CRISTIANI M, et al. Asymmetric Landau-Zener tunneling in a periodic potential[J]. Phys. Rev. Lett., 2003, 91: 230406.

[75] FLEISCHHAUER M, IMAMOGLU A, MARANGOS J P. Electromagnetically induced transparency: Optics in coherent media[J]. Rev. Mod. Phys., 2005, 77(2): 633.

[76] BHATTACHARYYA K. Quantum decay and the Mandelstam-Tamm-energy inequality[J]. J.Phys. A: Mathematical and General, 1983, 16(13): 2993.

[77] TADDEI M M, ESCHER B M, DAVIDOVICH L, et al. Quantum speed limit for physical processes[J]. Phys. Rev. Lett., 2013, 110: 050402.

[78] HEGERFELDT G C. Driving at the quantum speed limit: Optimal control of a two-level system[J]. Phys. Rev. Lett., 2013, 111: 260501.

[79] HEGERFELDT G C. High-speed driving of a two-level system[J]. Phys. Rev. A, 2014, 90(3): 032110.

[80] MARVIAN I, LIDAR D A. Quantum speed limits for leakage and decoherence[J]. Phys. Rev. Lett., 2015, 115(21): 210402.

[81] VILLAMIZAR D V, DUZZIONI E I. Quantum speed limit for a relativistic electron in a uniform magnetic field[J]. Phys. Rev. A, 2015, 92: 042106.

[82] MARVIAN I, SPEKKENS R W, ZANARDI P. Quantum speed limits, coherence, and asymmetry[J]. Phys. Rev. A, 2016, 93(5): 052331.

[83] PIRES D P, CIANCIARUSO M, CÉLERI L C, et al. Generalized geometric quantum speed

limits[J]. Phys. Rev. X, 2016, 6(2): 021031.

[84] LLOYD S, MONTANGERO S. Information theoretical analysis of quantum optimal control[J]. Phys. Rev. Lett., 2014, 113(1): 010502.

[85] CARLINI A, HOSOYA A, KOIKE T, et al. Time-optimal quantum evolution[J]. Phys. Rev. Lett., 2006, 96(6): 060503.

[86] BAKSIC A, RIBEIRO H, CLERK A A. Speeding up adiabatic quantum state transfer by using dressed states[J]. Phys. Rev. Lett., 2016, 116(23): 230503.

[87] LIU J, WU B, NIU Q. Nonlinear evolution of quantum states in the adiabatic regime[J]. Phys. Rev. Lett., 2003, 90(17): 170404.

[88] VITANOV N V, SUOMINEN K A. Nonlinear level-crossing models[J]. Phys. Rev. A, 1999, 59: 4580-4588.

[89] DORIA P, CALARCO T, MONTANGERO S. Optimal control technique for many-body quantum dynamics[J]. Phys. Rev. Lett., 2011, 106(19): 190501.

[90] CAMPBELL S, DE CHIARA G, PATERNOSTRO M, et al. Shortcut to adiabaticity in the Lipkin-Meshkov-Glick model[J]. Phys. Rev. Lett., 2015, 114(17): 177206.

[91] BROUZOS I, STRELTSOV A I, NEGRETTI A, et al. Quantum speed limit and optimal control of many-boson dynamics[J]. Phys. Rev. A, 2015, 92(6): 062110.

[92] FU L B, LIU J. Quantum entanglement manifestation of transition to nonlinear self-trapping for Bose-Einstein condensates in a symmetric double well[J]. Phys. Rev. A, 2006, 74(6): 063614.

[93] CHIN C, GRIMM R, JULIENNE P, et al. Feshbach resonances in ultracold gases[J]. Rev. Mod. Phys., 2010, 82: 1225-1286.

[94] DOU F Q, LIU J, AND FU L B. Fast quantum driving in two-level systems with interaction and nonlinear sweep[J]. Phys. Rev. A, 2018, 98: 022102.

[95] DAEMS D, RUSCHHAUPT A, SUGNY D, et al. Robust quantum control by a single-shot shaped pulse[J]. Phys. Rev. Lett., 2013, 111(5): 050404.

[96] VITANOV N V, RANGELOV A A, SHORE B W, et al. Stimulated Raman adiabatic passage in physics, chemistry, and beyond[J]. Rev. Mod. Phys., 2017, 89(1): 015006.

[97] LEVITT M H. Symmetrical composite pulse sequences for NMR population inversion. I. Compensation of radiofrequency field inhomogeneity[J]. Journal of Magnetic Resonance (1969), 1982, 48(2): 234-264.

[98] RIEBE M, MONZ T, KIM K, et al. Deterministic entanglement swapping with an ion-trap quantum computer[J]. Nature Physics, 2008, 4(11): 839-842.

[99] PILTZ C, SCHARFENBERGER B, KHROMOVA A, et al. Protecting conditional quantum gates by robust dynamical decoupling[J]. Phys. Rev. Lett., 2013, 110(20): 200501.

[100] HILL C D. Robust controlled-NOT gates from almost any interaction[J]. Phys. Rev. Lett., 2007, 98(18): 180501.

[101] GENOV G T, VITANOV N V. Dynamical suppression of unwanted transitions in multistate quantum systems[J]. Phys. Rev. Lett., 2013, 110(13): 133002.

[102] DRIDI G, GUERIN S, HAKOBYAN V, et al. Ultrafast stimulated Raman parallel adiabatic passage by shaped pulses[J]. Phys. Rev. A, 2009, 80(4): 043408.

[103] IBÁÑEZ S, CHEN X, TORRONTEGUI E, et al. Multiple Schrödinger pictures and dynamics in shortcuts to adiabaticity[J]. Phys. Rev. Lett., 2012, 109(10): 100403.

[104] VITANOV N V. Arbitrarily accurate narrowband composite pulse sequences[J]. Phys. Rev. A, 2011, 84(6): 065404.

[105] KYOSEVA E, VITANOV N V. Arbitrarily accurate passband composite pulses for dynamical suppression of amplitude noise[J]. Phys. Rev. A, 2013, 88(6): 063410.

[106] SCHRAFT D, HALFMANN T, GENOV G T, et al. Experimental demonstration of composite adiabatic passage[J]. Phys. Rev. A, 2013, 88(6): 063406.

[107] 豆福全, 郑伟强. 两能级系统的高保真度布居数反转 [J]. 科学通报, 2016, 61(20): 2309-2315.

[108] DHAR L, ROGERS J A, NELSON K A. Time-resolved vibrational spectroscopy in the impulsive limit[J]. Chem. Rev., 1994, 94(1): 157-193.

[109] STOLOW A. Applications of wavepacket methodology[J]. Philosophical Transactions of the Royal Society of London A: Mathematical, Physical and Engineering Sciences, 1998, 356(1736): 345-362.

[110] SHAPIRO M, BRUMER P. Principles of the Quantum Control of Molecular Processes[M]. New Jersey: Wiley-Inter-Science, 2003.

[111] DIDDAMS S A, BERGQUIST J C, JEFFERTS S R, et al. Standards of time and frequency at the outset of the 21st century[J]. Science, 2004, 306(5700): 1318-1324.

[112] RUSCHHAUPT A, CHEN X, ALONSO D, et al. Optimally robust shortcuts to population inversion in two-level quantum systems[J]. New J. Phys., 2012, 14(9): 093040.

[113] LU X J, CHEN X, RUSCHHAUPT A, et al. Fast and robust population transfer in two-level quantum systems with dephasing noise and/or systematic frequency errors[J]. Phys. Rev. A, 2013, 88(3): 033406.

[114] BERRY M. Two-state quantum asymptotics[J]. Annals of the New York Academy of Sciences, 1995, 755(1): 303-317.

[115] DOU F Q, LI S C. Achieving high molecular conversion efficiency via a magnetic field pulse train[J]. The European Physical Journal B-Condensed Matter and Complex Systems, 2012, 85(6): 1-6.

[116] FU L B, LIU J. Adiabatic Berry phase in an atom–molecule conversion system[J]. Ann. Phys., 2010, 325(11): 2425-2434.

[117] ZHANG D J, YU X D, TONG D. Theorem on the existence of a nonzero energy gap in adiabatic quantum computation[J]. Phys. Rev. A, 2014, 90(4): 042321.

[118] VITANOV N V, GARRAWAY B M. Landau-Zener model: Effects of finite coupling duration[J]. Phys. Rev. A, 1996, 53: 4288-4304.

[119] BATEMAN J, FREEGARDE T. Fractional adiabatic passage in two-level systems: Mirrors and beam splitters for atomic interferometry[J]. Phys. Rev. A, 2007, 76(1): 013416.

[120] MIAO X, WERTZ E, COHEN M, et al. Strong optical forces from adiabatic rapid passage[J].

Phys. Rev. A, 2007, 75(1): 011402.

[121] LU T, MIAO X, METCALF H. Nonadiabatic transitions in finite-time adiabatic rapid passage[J]. Phys. Rev. A, 2007, 75(6): 063422.

[122] LU T. Population inversion by chirped pulses[J]. Phys. Rev. A, 2011, 84(3): 033411.

[123] HU J Y, MAO T F, DOU F Q, et al. Application of the composite adiabatic passage technique in the Landau-Zener model with harmonic interaction modulation[J]. Acta Phy. Sin., 2013, 62(17): 170303.

[124] TONG D, SINGH K, KWEK L C, et al. Sufficiency criterion for the validity of the adiabatic approximation[J]. Phys. Rev. Lett., 2007, 98(15): 150402.

[125] LI S C, FU L B, DUAN W S, et al. Nonlinear Ramsey interferometry with Rosen-Zener pulses on a two-component Bose-Einstein condensate[J]. Phys. Rev. A, 2008, 78(6): 063621.

[126] LIU J, WU B, FU L, et al. Quantum step heights in hysteresis loops of molecular magnets[J]. Phys. Rev. B, 2002, 65(22): 224401.

[127] SMERZI A, FANTONI S, GIOVANAZZI S, et al. Quantum coherent atomic tunneling between two trapped Bose-Einstein condensates[J]. Phys. Rev. Lett., 1997, 79(25): 4950.

[128] ALBIEZ M, GATI R, FÖLLING J, et al. Direct observation of tunneling and nonlinear self-trapping in a single bosonic Josephson junction[J]. Phys. Rev. Lett., 2005, 95(1): 010402.

[129] KHOMERIKI R. Nonlinear Landau-Zener tunneling in coupled waveguide arrays[J]. Phys. Rev. A, 2010, 82: 013839.

[130] CRISTIANI M, MORSCH O, MÜLLER J, et al. Experimental properties of Bose-Einstein condensates in one-dimensional optical lattices: Bloch oscillations, Landau-Zener tunneling, and mean-field effects[J]. Phys. Rev. A, 2002, 65(6): 063612.

第 5 章　超冷原子–分子转化动力学

本章主要介绍超冷原子–分子转化动力学,包括超冷原子–双原子分子的转化以及超冷原子–多原子分子的转化问题,提出了磁场脉冲链技术和广义受激拉曼绝热通道技术,分别从理论上实现了高效超冷原子–双原子分子的转化和超冷原子–多原子分子的转化,讨论了外场参数以及外场形式原子–分子转化效率的影响.

5.1　超冷双原子分子的产生: 磁场脉冲链技术的量子操控

本节通过包含粒子间相互作用的超冷玻色系统, 研究了该系统中超冷原子–双原子分子的转化问题. 通过在 Feshbach 共振附近设计一个磁场脉冲链, 实现了一种稳定、高效地产生超冷双原子分子的方法. 分子转化效率和每个磁场脉冲的持续时间紧密相关, 适当地调节各个磁场脉冲的持续时间, 分子转化效率可高达 100%, 也得到了这些磁场脉冲持续时间的解析表达式. 进一步研究表明, 分子的转化效率极其敏感于第二个磁场脉冲, 敏感于第一个磁场脉冲, 而鲁棒于第三个磁场脉冲. 同时也讨论了粒子间相互作用对原子–分子转化过程的影响.

超冷分子的研究是过去十几年里最活跃的研究领域之一, 并持续吸引人们的注意 [1-3]. 目前, 被广泛应用于产生超冷分子的两种主要技术是光缔合和磁场 Feshbach 共振. 除此之外, 一种组合光缔合和 Feshbach 共振的方法, 也称为优化 Feshbach 的光缔合技术 [4-6], 也被用于产生超冷分子.

当双原子分子态和两个散射原子态之间的能级差为零时, 会发生 Feshbach 共振. 利用 Feshbach 共振已经发展了许多产生超冷分子的技术, 形成的分子通常称为 Feshbach 分子, 这些方法主要应用于随时间变化的磁场. 为了产生超冷 Feshbach 分子, 人们普遍采用外部磁场扫描经过 Feshbach 共振, 即第 2 章中提到的被称作 Feshbach 扫描的技术 [7]. 原子转化到 Feshbach 分子的实验是在 ^{85}Rb 原子的玻色–爱因斯坦凝聚中首次被观察到的, 该实验运用 Ramsey 类型 [8] 快速随时间变化的磁场 (几十微秒). 然而, 到目前为止最流行的分子转化技术所采用的磁场是慢速扫描通过 Feshbach 共振. 另外一种产生超冷 Feshbach 分子的方法是基于调制的磁场 [9,10], 这种振荡磁场诱发两个碰撞的原子形成束缚分子态, 同时提高了分子转化效率. 第三种原子–分子转化技术是在 Feshbach 共振附近使用提高三体碰撞重组率的方式有效地缔合费米原子成为分子 [11].

5.1 超冷双原子分子的产生: 磁场脉冲链技术的量子操控

以上方法中, Feshbach 扫描技术已经成为产生超冷分子的标准技术, 并成为研究 Feshbach 分子动力学和相互作用性质的有力工具. 该方法已经用于将玻色和费米原子转化为超冷同核、异核分子[7]. 然而, 在目前的实验里, 这种方法的转化效率相对较低, 为 4%[12]~80%[13], 而 50%~70% 的原子在扫描过程中由于诱导发热而损失[14]. 为了提高转化效率, 在理论和实验上, 原子 BECs 通过磁场脉冲方式操控已经有所报道[8,14–17], 特别是通过采用一系列磁场脉冲, 转化效率可达到 100%[17]. 然而, 这种方法的分子布居数仍然在原子与分子之间振荡, 即随着时间 t 的推移, 分子很快又离解成自由原子. 同时为了简单起见, 该方法中粒子之间的相互作用, 包括原子与原子、原子与分子、分子与分子之间的相互作用并未考虑.

本节修正了文献 [17] 的方法, 从理论上研究了超冷玻色系统中超冷原子–分子的转化问题. 作为文献 [17] 方法的扩展, 模型考虑了粒子间非线性相互作用, 并且在原来脉冲基础上又增加了一些磁场脉冲, 构造了一个磁场脉冲链. 这种修正的方法有效地抑制了分子布居数的振荡, 形成一种稳定的、转化效率为 100% 的原子–分子转化方法. 每个磁场脉冲的持续时间也能够通过解析获得. 同时发现, 分子转化效率高度敏感于第二个磁场脉冲, 而鲁棒于第三个磁场脉冲, 该脉冲对应于一个宽的 100% 转化效率的范围. 另外, 也讨论了粒子间相互作用对转化过程的影响[18].

5.1.1 平均场模型和拉比振荡

两模模型近似下, 包括原子–原子、原子–分子、分子–分子相互作用的双通道模型如下[19,20]:

$$\hat{H} = \epsilon_a \hat{a}^\dagger \hat{a} + \epsilon_b \hat{b}^\dagger \hat{b} + \frac{g}{\sqrt{V}}(\hat{a}^\dagger \hat{a}^\dagger \hat{b} + \text{H.c.})$$
$$+ \frac{u_a}{V}\hat{a}^\dagger \hat{a}^\dagger \hat{a} \hat{a} + \frac{u_b}{V}\hat{b}^\dagger \hat{b}^\dagger \hat{b} \hat{b} + \frac{u_{ab}}{V}\hat{a}^\dagger \hat{a} \hat{b}^\dagger \hat{b}, \tag{5.1}$$

其中, 算符 $\hat{a}(\hat{a}^\dagger)$ 和 $\hat{b}(\hat{b}^\dagger)$ 分别是原子和分子的湮灭 (产生) 算符; $g = \sqrt{2\pi\hbar^2 a_{bg}\Delta B \mu_{co}/m}$[21,22] 指原子和分子之间的 Feshbach 耦合强度, 其中 a_{bg} 是背景散射长度, m 是原子质量, ΔB 代表共振宽度, 并且 μ_{co} 是分子和一对分离原子之间的磁矩差; u_a 和 u_b 分别指原子–原子与分子–分子之间的相互作用, 而 u_{ab} 指原子–分子之间的相互作用, 并且 $u_a = 2\pi\hbar^2 a_{bg}/m$, $u_b = \pi\hbar^2 a_{bb}/m$, $u_{ab} = 3\pi\hbar^2 a_{ab}/m$[21,23], 这里 a_{bb} 和 a_{ab} 分别定义为分子–分子以及原子–分子相互作用的散射长度; ϵ_a 和 ϵ_b 分别指原子态和分子态的能量. 在实验里, 外部磁场 $B(t)$ 扫描经过 Feshbach 共振点 B_0, 这样就有 $2\epsilon_a - \epsilon_b = \mu_{co}[B(t) - B_0]$. 引入参数 V 表示俘获粒子的量子化的体积, $n = N/V$ 是初始玻色原子的平均密度. 系统的整个粒子总数 $N = \hat{a}^\dagger \hat{a} + 2\hat{b}^\dagger \hat{b}$ 是一个常数. 这种两模模型已经成功地用于解释通过线性扫描磁场[24,25] 和正弦振荡场[10] 下 Feshbach 原子–分子转化的相关实验.

在目前的实验中, 一般粒子数非常大, 可以采用平均场近似. 在这种近似下, 总粒子数趋于无穷而密度固定不变, 在平均值附近的量子和热涨落可以忽略不计[26]. 此时, 场算符 $\hat{a}$ 和 $\hat{b}$ 可用序参数场中的 c 数代替, 即 $\hat{a} \sim \langle \hat{a} \rangle = \sqrt{N}\sqrt{p_a}e^{i\theta_a}$, $\hat{b} \sim \langle \hat{b} \rangle = \sqrt{N}\sqrt{p_b}e^{i\theta_b}$, 其中 p_a 和 p_b 分别指原子和分子布居数, θ_a 和 θ_b 是对应的相位. 总粒子数守恒, $|\langle\hat{a}\rangle|^2 + 2|\langle\hat{b}\rangle|^2 = N$ 使得 $p_a + 2p_b = 1$. 下面引进正则变换, $p = p_b$, $\theta = 2\theta_a - \theta_b$, 这里 θ 指两模的相对相位. 此时, 湮灭算符 $\hat{a}$ 和 $\hat{b}$ 满足的 Heisenberg 运动方程就会等价于下面的经典哈密顿量:

$$H_{cl} = \Delta p + cp^2 + G(1-2p)\sqrt{p}\cos\theta, \tag{5.2}$$

其中, $\Delta = -(2\epsilon_a - \epsilon_b + 4nu_a - nu_{ab})$; $c = n(4u_a + u_b - 2u_{ab})$; $G = 2g\sqrt{n}$. 为简单起见, 在下面的计算和所有的数值模拟中, 耦合强度 G 作为能量尺度, 并取 $G = 1$.

经典哈密顿量 (5.2) 描述系统的总能量, 正则变量 p 和 θ 满足 $\dot{\theta} = \partial H_{cl}/\partial p$, $\dot{p} = -\partial H_{cl}/\partial \theta$, 即

$$\dot{\theta} = \Delta + 2cp - \frac{G(6p-1)}{2\sqrt{p}}\cos\theta, \tag{5.3}$$

$$\dot{p} = G(1-2p)\sqrt{p}\sin\theta. \tag{5.4}$$

上述经典哈密顿量的不动点可以通过解方程组 $\dot{\theta} = 0$ 和 $\dot{p} = 0$ 得到. $p = 1/2$ 对应于纯分子态, 它不依赖于相对相位 (此时 θ 没有定义). 图 5.1 显示了一些典型参数下经典哈密顿量 (5.2) 的相空间结构, 图中圆圈代表不动点. 图 5.1 (c) 中, 在线 $p = 1/2$ 上的两个不动点实际上是同一个不动点. 不动点周围的轨道代表原子态和分子态之间的 Rabi 振荡, 对应于不同的初始条件.

Rabi 振荡的周期可以通过经典能量 (5.2) 守恒很容易获得

$$T_R = 2\int_{p_0}^{p_m} \frac{dp}{\sqrt{G^2(1-2p)^2 p - (H_0 - \Delta p - cp^2)^2}}, \tag{5.5}$$

其中, H_0 是初始态能量; p_0 和 p_m 分别对应着能到达的分子布居数 p 的最大值和最小值, 它可以由下述方程得到:

$$G^2(1-2p)^2 p - (H_0 - \Delta p - cp^2)^2 = 0. \tag{5.6}$$

显而易见, 当 $H_0 = 0$ 时 $p_0 = 0$, 表示初始态是一个纯原子态. 系统在初始纯原子态下原子态和分子态的振荡的时间演化如图 5.2(a) 所示, Rabi 振荡的周期 T_R 随 Δ 和 c 的演化如图 5.2(b) 和 (c) 所示. 初始的相对相位选为 $\theta(0) = 0$. 从图 5.2 可以看到: ① Rabi 振荡的振幅和周期由于粒子之间的非线性相互作用而发生了极大的改变, ② Rabi 的周期在 Δ 和 c 取一些特殊值时趋于发散.

5.1 超冷双原子分子的产生: 磁场脉冲链技术的量子操控

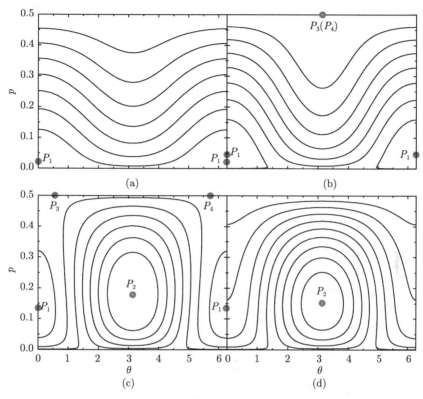

图 5.1 哈密顿量 (5.2) 的相空间结构

(a) $\Delta = -3.0, c = 0.3$; (b) $\Delta = -1.7142, c = 0.3$; (c) $\Delta = -0.8, c = 2.0$; (d) $\Delta = -0.8, c = 3.0$

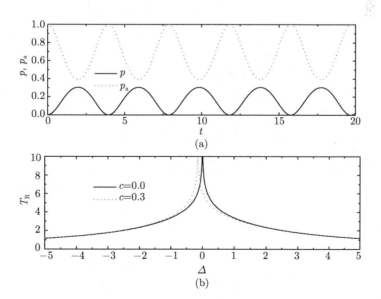

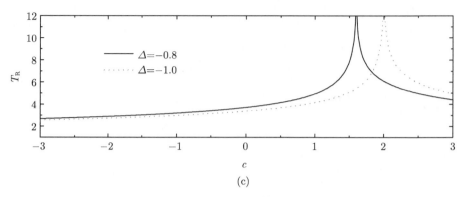

图 5.2　Rabi 振荡

(a) $\Delta = -0.8, c = 0.3$ 时 Rabi 振荡的时间演化; (b) Rabi 振荡的周期随 Δ 的变化; (c) Rabi 振荡的周期随非线性参数 c 的变化

5.1.2　原子-分子转化的磁场脉冲链技术

本节中推广了文献 [17] 中的技术, 通过设计一个磁场脉冲链实现稳定高效的原子-分子转化. 原子和分子布居数随时间的演化如图 5.3(a) 所示. 参数 Δ 可以通过改变随时间变化的磁场 $B(t)$ 来调节. 对 Δ 调节所采用的方式如图 5.3(b) 所示, 图中竖直虚线将时间依赖的四段磁场脉冲分成对应的 I、II、III、IV 四个演化过程, 即该磁场脉冲链由四段常磁场组成, 磁场的值要么临近 Feshbach 共振, 要么远离 Feshbach 共振. $T_i (i = 1, 2, 3, 4)$ 表示每个脉冲的持续时间. 当 $0 \leqslant t < T_1$ 和 $(T_1 + T_2) \leqslant t < (T_1 + T_2 + T_3)$ 时, 磁场脉冲对应于近共振的值 Δ_1, 而当 $T_1 \leqslant t < (T_1 + T_2)$ 和 $t \geqslant (T_1 + T_2 + T_3)$ 时, 对应于远共振值 Δ_2. 为了在第一个磁场脉冲结束时分子达到最大的转化, 选择 $T_1 = T_R/2$. 在 $T_1 \leqslant t < (T_1 + T_2)$ 过程中, 磁场远离共振值, 演化过程中伴随着小振幅的快速振荡. 选择一个特殊的值 T_2, 以确保当 $t > (T_1 + T_2)$ 时, 系统经历一个大振幅的慢 Rabi 振荡 (服从能量守恒), 其最大值能接近极限值——完全转化, 即 $p = 0.5$. 为了进一步确保分子转化效率保持在 100%, 在 $t \geqslant (T_1 + T_2 + T_3)$ 时设置了第四个磁场脉冲.

每个脉冲的持续时间可以通过经典能量 (5.2) 来计算, 得到下面的解析表达式:

$$T_1 = \frac{T_R}{2} = \int_0^{p_m} \frac{\mathrm{d}p}{\sqrt{G^2(1-2p)^2 p - (\Delta_1 p + cp^2)^2}}, \tag{5.7}$$

$$T_2 = n \oint \left| \frac{\partial p}{\partial H_{cl}} \right| \mathrm{d}\theta + \int_0^{\theta_{\max}} \left| \frac{\partial p}{\partial H_{cl}} \right| \mathrm{d}\theta$$

5.1 超冷双原子分子的产生: 磁场脉冲链技术的量子操控

$$\simeq n \int_0^{2\pi} \left| \frac{1}{-\Delta_2 - 2cp_\mathrm{m} + \frac{G(6p_\mathrm{m}-1)}{2\sqrt{p_\mathrm{m}}}\cos\theta} \right| \mathrm{d}\theta$$

$$+ \int_0^{\theta_\mathrm{m}} \left| \frac{1}{-\Delta_2 - 2cp_\mathrm{m} + \frac{G(6p_\mathrm{m}-1)}{2\sqrt{p_\mathrm{m}}}\cos\theta} \right| \mathrm{d}\theta, \tag{5.8}$$

$$T_3 = \int_{p_\mathrm{m}}^{0.5} \frac{\mathrm{d}p}{\sqrt{G^2(1-2p)^2 p - (H_0 - \Delta_1 p - cp^2)^2}}, \tag{5.9}$$

其中, p_m 满足方程 (5.6), n 是一个整数, 定义为第二个脉冲的振荡周期个数; $\theta_{\max} = 2\pi - \arccos\left\{(4p_\mathrm{m}^2 - 2\Delta_1 + 4p_\mathrm{m}\Delta_1 - c)/[4G\sqrt{p_\mathrm{m}}(2p_\mathrm{m}-1)]\right\}$, 指在第二个磁场脉冲结

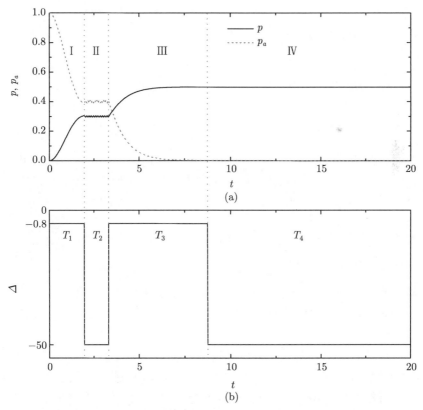

图 5.3 原子和分子布居数随时间变化和磁场脉冲链调节方式示意图

(a) 原子和分子布居数随时间的演化; (b) 磁场脉冲链调节方式 Δ 随时间变化的示意图

束时对应的最大相位; $H_0 = \Delta_1 p_{\mathrm{m}} + c p_{\mathrm{m}}^2 + G(1-2p_{\mathrm{m}})\sqrt{p_{\mathrm{m}}}\cos(\theta_{\mathrm{m}})$ 代表在第三个脉冲其间的系统能量. 当保持外部磁场不变时, 每个脉冲的持续时间强烈地依赖于非线性相互作用, 特别是第一和第三个磁场脉冲的持续时间 T_1 和 T_3, 这些结果如图 5.4 所示, 图中其他参数值分别为 $\Delta_1 = -0.8, \Delta_2 = -50, n = 10$. 在弱非线性相互作用范围, 持续时间 T_1 逐渐增加而 T_3 逐渐降低. 相反地, 在强非线性相互作用范围, T_1 降低而 T_3 增加, 当 c 取一些特殊值时, 它们的值却趋于发散.

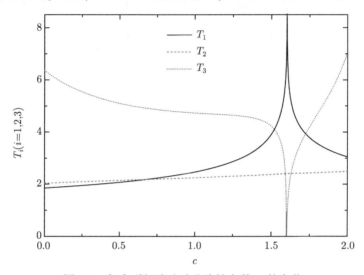

图 5.4 每个磁场脉冲随非线性参数 c 的变化

在所有数值模拟中, 取 $\Delta_1 = -0.8, \Delta_2 = -50, n = 10$. 关于非线性相互作用的取值, 考虑 MIT 在 ^{23}Na 凝聚中的实验参数, 凝聚的平均密度为 $n \sim 10^{15}\mathrm{cm}^{-3}$, 在磁场 $B_0 = 853\mathrm{G}$ 处发生 Feshbach 共振时, $\Delta B = 0.0025\mathrm{G}, a_{bg} = 3.33\mathrm{nm}, \mu_{co} = 3.52 \times 10^{-23}\mathrm{J}\cdot\mathrm{T}^{-1}$ (可文献 [7], [21], [27]), 这样获得 $c/G \approx 0.5\sqrt{\dfrac{\pi\hbar^2 n a_{bg}}{m\Delta B \mu_{co}}} \approx 0.3$, 同时时间尺度为 $7.196 \times 10^{-6}\mathrm{s}$.

这样得到 $T_1 \simeq 1.97$, 在第一个脉冲结束时对应的分子转化效率的最大值是 $p_{\mathrm{m}} \simeq 0.31$, 系统振荡 10 个周期其间的 $T_2 \simeq 1.345$. 在 $t = 8.75$ 时具有最大的分子转化效率 $p \simeq 0.499$, 如图 5.3 所示.

分子转化效率随每个磁场脉冲持续时间的变化情况如图 5.5 所示. 可以看到图 5.5(b) 中的振荡比图 5.5(a) 和 (c) 中都快, 也就是说, 分子转化效率高度敏感于第二个脉冲持续时间, 而鲁棒于第三个脉冲持续时间, 存在一个宽的 100% 转化效率的时间窗口. 之所以如此, 是由于在第二个磁场脉冲其间, Rabi 振荡的周期比第一个和第三个脉冲的周期要小得多.

5.1 超冷双原子分子的产生: 磁场脉冲链技术的量子操控

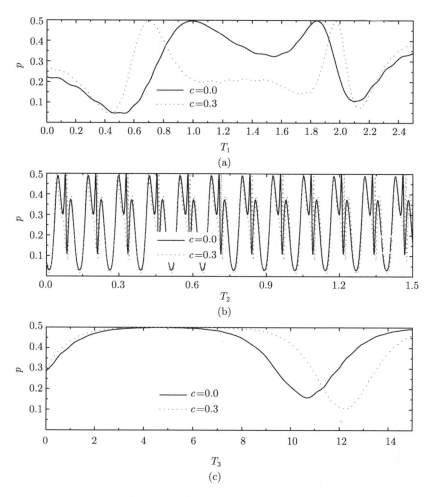

图 5.5 分子转化效率随每个磁场脉冲持续时间的变化图
(a) $p-T_1$; (b) $p-T_2$; (c) $p-T_3$

为了进一步理解如图 5.2 中高效稳定分子转化效率磁场操控方式的物理机制, 借助哈密顿量 (5.2) 在 p 和 θ 的相空间结构来解释这种演化行为, 如图 5.6 所示, 其中箭头表示演化方向, 在相空间中 $A(B,C)$ 和 $A'(B',C')$ 实际上是相同的点. 演化过程 I、II、III和IV分别对应着四个随时间变化的磁场脉冲. 第一个磁场脉冲的磁场值在近 Feshbach 共振, 磁场扫描其间系统做 Rabi 振荡, 服从能量守恒定律. 在 $t=T_1$ 时, 磁场突然改变到远离共振的值, 这个时刻, 相空间的瞬时位置不变 (A 对应于 A'). 磁场改变之后, 系统的能量也会随之发生改变. 在第二个磁场脉冲其间, 系统经历了一个小振幅的快速振荡, 当 $t=T_1+T_2$ 时, 系统演化到位置 B'. 随后, 磁场的值又返回近共振的值, 系统会位于一个新的位置 B (对应于 B'), 这时系统不再跟随原来的第一个磁场脉冲引起的 Rabi 振荡的轨道, 而是进入另一个常能量的轨道. 在磁场脉冲Ⅲ其间, 系统伴随着一个大振幅的慢速 Rabi 振荡, 此时的分子转化效率的最大值几乎达到极限值, $p = 0.5$, 实现了完全的转化. 在 $t = T_1 + T_2 + T_3$ 时刻, 系统演化到图 5.6(a) 中的位置 C. 和文献 [17] 中图 3 不同的是, 当系统位置在 C' (对应于 C) 点时, 将磁场值又返回到和第二脉冲时一样的值, 即远共振的值. 有意义的是, 在整个磁场Ⅲ过程中, 系统将始终保持同样一个值, 即最大的转化效率.

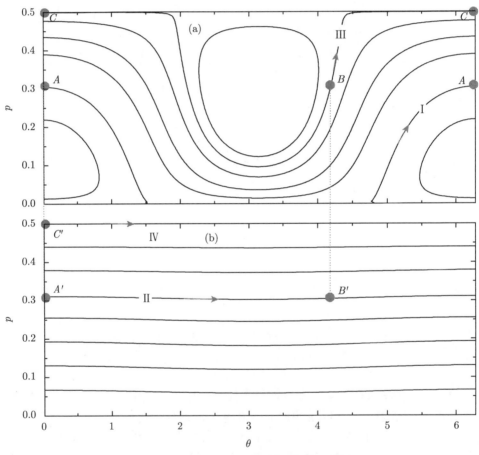

图 5.6　哈密顿量 (5.2) 在不同参数下相图

(a) $\Delta = -0.8$; (b) $\Delta = -50$

5.1.3　结论和讨论

本节研究了包含原子–原子、原子–分子和分子–分子之间相互作用的玻色系统中原子到双原子分子的转化问题; 讨论了原子态和分子态之间的 Rabi 振荡, 发现非线性相互作用将明显影响 Rabi 振荡的周期. 通过设计一个磁场脉冲链, 适当地调控每个磁场脉冲的持续时间, 能实现原子到分子的完全转化, 而这些磁场脉冲的持续时间能够解析得到. 讨论了粒子间非线性相互作用对每个磁场脉冲持续时间的影响. 研究表明, 分子的转化效率高度敏感于第二个磁场脉冲持续时间, 而鲁棒于第一和第三个磁场脉冲持续时间, 特别是第三个磁场持续时间, 存在一个 100% 转化效率的宽窗口. 最后, 借助于经典哈密顿量的相图, 基于能量守恒定律, 给出了一些理论解释.

这里所采取的技术优于先前的方法, 这是由于该方法提供了一种稳定的、高效的分子转化技术, 同时不要求磁场穿过 Feshbach 共振点, 以致避免了系统的热损失和粒子丢失 [17]. 过程中粒子的主要损失是三体重组损失. 然而, 在实验中三体损失的时间尺度的量级是毫秒 [28], 比该技术中近共振时磁场持续时间 (典型的时间尺度为 7.196×10^{-6} s, 微秒量级) 要长. 事实上, 磁场的扫描率在实验上是有限的, 我们给出它的一个下限, 即, $k_B T|B_2 - B_1|/(2\pi\hbar)$ [25] (磁场 B_1 和 B_2 分别对应于前面无量纲化的参数 Δ_1 和 Δ_2).

在实际的实验中, 热粒子存在是不可避免的. 两模模型对非凝聚的原子和分子气体是一种近似描述, 它仅适用于当热粒子的能量分布 (由 $k_B T$ 表示, k_B 是玻尔兹曼常量, T 是温度) 远远小于等效 Feshbach 共振宽度 $g\sqrt{n}$ 的情况. 另外, 分子玻色-爱因斯坦凝聚的单模描述也仅在短时间演化符合得好 [10], 对于长时间的演化, 热粒子散射偏离单模平均场将导致其正比于热原子团温度的相扩散 [29]. 另外, 也应该注意到, 磁场脉冲的扫描会导致产生一个附加的成分, 它可以作为一个由关联原子对组成的非凝聚的 burst 原子 [30,31]. 为了考虑这种效应, 本章中的两模模型还应该适当扩展, 将 burst 原子作为一种新的成分考虑进去.

5.2 超冷 N 体 Efimov 多聚物分子的形成: 广义受激拉曼绝热通道技术的量子操控

5.1 节研究了超冷原子到双原子分子的转化, 本节仍然考虑超冷玻色系统, 研究超冷原子到多聚物分子的转化问题. 首先考虑超冷五聚物分子的产生, 接着对该方法进行拓展, 研究超冷 N 体 Efimov 多聚物分子的形成. 通过一种广义的受激拉曼绝热通道技术, 建立平均场模型, 获得系统的 CPT 解, 对多聚物分子, 其 CPT 解满足一个普适的代数方程. 利用线性不稳定分析方法, 分析暗态的线性不稳定性, 并通过定义合适的保真度, 研究系统的绝热性. 同时也讨论粒子间相互作用、多聚物分子的原子数目以及外场参数对原子-多聚物分子转化过程的影响 [32,33].

超冷分子气体的研究叩响了研究新的物理现象的大门, 有着十分广泛的应用, 如超冷分子能用于检测基本对称性[34]、精密光谱测量[35]、量子信息过程[36,37], 以及超冷化学[38,39]等领域. 光缔合 (PA) 和磁场 Feshbach 共振 (FR) 是超冷原子形成两体相互作用势下超冷分子束缚态的两种主要方法. 然而, 由这两种方法形成的双原子分子通常是松散的束缚态并且极其不稳定. 因此, 人们不得不采用受激拉曼绝热通道 (STIRAP) 技术将超冷原子绝热转化为紧束缚的基态分子. 人们首先提议的是将非凝聚的原子缔合成稳定分子, Feshbach 辅助的受激拉曼绝热通道技术 (FR-aided STIRAP)[40] 被认为是一种比纯 STIRAP 技术转化原子凝聚体成分子气体更为有效的方法. 受激拉曼绝热通道技术的成功主要依赖于存在相干布居数俘获 (CPT) 态[41], 即暗态, 系统应该绝热地跟随这个暗态. 对线性系统而言, 这样一个条件可以通过适当地选择激光频率而满足. 然而, 对于包含粒子间相互作用的非线性系统, 当布居数从原子态向分子态转化时, 双光子共振条件会发生动力学改变. 这使得绝热跟随 CPT 态更加困难, 从而导致低的原子-分子转化效率. 为了处理这样一个问题, 第 2 章中提到广义的 STIRAP 技术[42] 被采用了, 它通过一个啁啾耦合场去补偿粒子间相互作用的非线性效应, 同时高效地产生大量深束缚态超冷分子.

尽管人们对超冷分子的研究热情一直未减并不断升温[43,44], 然而, 该领域大部分的实验和理论工作主要聚焦于双原子分子[1,7,45–49]. 超冷分子领域另一个有趣的课题就是获得超冷复杂分子, 目前研究的一个重要进展就是扩展研究超冷双原子分子的技术到超冷多原子分子[50,51]. 冷却和俘获多原子分子极大地丰富了分子物理领域, 同时也用于研究复杂化学反应 (包括超冷化学和超化学)[52]、分子退相干[53]、精密测量[35] 以及分子光学[47,54].

形成超冷多原子分子在原则上是可能的, 要么从室温直接冷却原来的分子, 要么从原子和更小的分子通过非直接的方法缔合成多原子分子. 最近, 实验上报道了一个直接冷却多原子分子的方法, 称作 Sisyphus 冷却[55]. 该技术能将氟代甲烷 CH_3F 分子冷却到 29mK, 甚至温度还能低于 1mK, 它代表了一个新的冷却和俘获分子的有用方法, 也是多原子分子气体形成中的一个惊奇应用[56].

一种潜在的利用非直接方法形成超冷多原子分子的可行方法是, 紧接着超冷原子中随磁俘获技术和之后的共同冷却方法[57–61] 以及低温缓冲气体冷却[62] 或者 Stark 减速[63]. 然而, 这些方法 (如 Stark 减速) 仍然不能到达超冷温度范围.

另一个有希望产生超冷多原子分子的方法和磁场调节的 Feshbach 共振有关. 借助 Feshbach 共振和光缔合的广义的受激拉曼绝热通道技术[42], 不仅被用于产生大量的深束缚态超冷双原子分子, 同时也被用于形成同核和异核超冷三原子分子[64,65].

5.2 超冷 N 体 Efimov 多聚物分子的形成：广义受激拉曼绝热通道技术的量子操控

从少体物理角度而言，在临近 Feshbach 共振中心，出现了三体 Efimov 共振现象，三体 Efimov 共振 (ER) 分子早在 19 世纪 70 年代就被预言[66]，直到 2005 年才首次在超冷气体中被观察到[67]。Efimov 共振的实现不仅证实了存在弱束缚的三体态，还打开了实验和理论上探索引起人们极大兴趣的少体量子系统的大门[68]。在过去的几年里，Efimov 物理领域已经在超冷原子气体中发现了许多令人兴奋的现象[38,69,70]。将玻色气体加载在光晶格上已经被提议用来产生三体 Efimov 分子[71]。Efimov 共振也很快推广到四体系统[72,73]，对四个全同的玻色子而言[74-77]，人们发现 Efimov 三体态和四体态总是一起出现，每一个三体态能级会缔合两个四体态。实验上，四体态也已经在超冷铯原子气体中实现了[78]。最近的理论研究已经将 Efimov 物理做了进一步的推进，其中 $N = 5, 6, 7$ 体或者更多体团簇态 (N 体 Borromean) 能被形成，甚至在不存在束缚子系统时也毫不影响团簇态的形成[76,79-81]。人们也研究了 $N = 40$ 个原子的团簇态，计算了连续的 N 个玻色系统穿过对应的原子阈值时的散射长度的值[82]。

目前，借助于平均场理论，广义的受激拉曼绝热通道技术已经用于获得同核[83]和异核[84]四聚物分子。很自然地，人们不禁会问：借助三体和多体 Efimov 共振和光缔合技术，能否构建一种广义的受激拉曼绝热通道技术去实现更复杂的超冷 N 体多原子分子？

本节将探讨这方面的问题，首先形成了超冷五聚物分子，然后对此进一步推广，形成了稳定的超冷 N 体多聚物分子。相应地，本部分也分为三节，第 5.2.1 节为超冷五聚物分子的形成，第 5.2.2 节为超冷 N 体 Efimov 多聚物分子的形成，第 5.2.3 节为小结。

5.2.1 超冷五聚物分子的形成

本节的目的在于应用广义的受激拉曼绝热通道技术，从理论上示范该技术可以实现超冷同核和异核五聚物分子的转化。其基本过程是：首先从超冷原子中产生四聚物 A_4，然后将它和另一个原子通过光缔合形式耦合成一个束缚的五聚物 A_5 或 A_4B。在整个过程中，获得一个相干原子–分子暗态至关重要，它被用来阻止四聚物分子布居数。得到原子-五聚物分子暗态解之后，主要聚焦于暗态的稳定性和绝热性。为了选择合适的外场参数值，从而有效地实现原子到五聚物分子的转化，讨论了单光子失谐、Rabi 脉冲的强度和宽度对转化效率的影响。

1. 平均场模型和 CPT 态

本节研究的模型首先由超冷原子耦合形成四聚物分子，然后和另一个原子通过光缔合形成五聚物分子。这个相干转化过程可以理解为一个抽象的三能级模型，自由原子态、高激发态四聚物分子态和五聚物分子态形成抽象的三能级系统。因此，该过程可以使用受激拉曼绝热通道技术.

首先考虑产生同核五聚物分子的情形. λ' 定义为原子–五聚物分子耦合强度, 对应的失谐是 δ. 光缔合的激光 Rabi 频率为 Ω', 其失谐为 Δ, 相互作用绘景下, 描述该系统的哈密顿量为

$$\hat{H} = -\hbar \int \mathrm{d}\boldsymbol{r} \Big\{ \sum_{i,j} \chi'_{i,j} \hat{\psi}_i^\dagger(\boldsymbol{r}) \hat{\psi}_j^\dagger(\boldsymbol{r}) \hat{\psi}_j(\boldsymbol{r}) \hat{\psi}_i(\boldsymbol{r}) + \delta \hat{\psi}_t^\dagger(\boldsymbol{r}) \hat{\psi}_t(\boldsymbol{r}) \\
+ \lambda' [\hat{\psi}_t^\dagger(\boldsymbol{r}) \hat{\psi}_a(\boldsymbol{r}) \hat{\psi}_a(\boldsymbol{r}) \hat{\psi}_a(\boldsymbol{r}) \hat{\psi}_a(\boldsymbol{r}) + \text{H.c.}] \\
+ (\Delta + \delta) \hat{\psi}_p^\dagger(\boldsymbol{r}) \hat{\psi}_p(\boldsymbol{r}) - \Omega' [\hat{\psi}_p^\dagger(\boldsymbol{r}) \hat{\psi}_t(\boldsymbol{r}) \hat{\psi}_a(\boldsymbol{r}) + \text{H.c.}] \Big\}, \tag{5.10}$$

其中, $\hat{\psi}_i$ 和 $\hat{\psi}_i^\dagger$ 表示湮灭和产生算符; $\chi'_{i,j}$ 代表两体相互作用, 下标 $i,j = a, t, p$ 分别代表原子态、四聚物分子态和五聚物分子态.

实验中粒子数足够大, 考虑的又是超冷低温系统, 因此可以采用平均场理论. 在这种近似下, 粒子数趋于无穷而密度保持固定, 在平均值附近的量子和热涨落可以忽略不计 [26]. 此时, 场算符 $\hat{\psi}_i$ 和 $\hat{\psi}_i^\dagger$ 就能够用序参数场中的 c 数 $\sqrt{n}\psi_i$ 和 $\sqrt{n}\psi_i^*$ 代替, 其中 n 是总粒子数密度. 湮灭算符 $\hat{\psi}_i$ 满足的 Heisenberg 方程就可以写作:

$$\begin{cases} \dfrac{\mathrm{d}\psi_a}{\mathrm{d}t} = 2\mathrm{i} \sum_j \chi_{aj} |\psi_j|^2 \psi_a + 4\mathrm{i}\lambda \psi_t \psi_a^{*3} - \mathrm{i}\Omega \psi_p \psi_t^*, \\ \dfrac{\mathrm{d}\psi_t}{\mathrm{d}t} = 2\mathrm{i} \sum_j \chi_{tj} |\psi_j|^2 \psi_t + (\mathrm{i}\delta - \gamma)\psi_t + \mathrm{i}\lambda \psi_a^4 - \mathrm{i}\Omega \psi_p \psi_a^*, \\ \dfrac{\mathrm{d}\psi_p}{\mathrm{d}t} = 2\mathrm{i} \sum_j \chi_{pj} |\psi_j|^2 \psi_p + \mathrm{i}(\Delta + \delta)\psi_p - \mathrm{i}\Omega \psi_t \psi_a, \end{cases} \tag{5.11}$$

其中, $\chi_{ij} = n\chi'_{ij}$; $\lambda = n\sqrt{n}\lambda'$; $\Omega = \sqrt{n}\Omega'$ 是重正化的量, 正比于 γ 的项, 用来模拟中间四聚物分子态上粒子的损失.

现在求方程组 (5.11) 满足 $|\psi_t^0| = 0$ 的稳定的 CPT 态解. 首先做如下稳态变换:

$$\begin{cases} \psi_a = |\psi_a^0| \exp\left[\mathrm{i}(\theta_a - \mu_a t)\right], \\ \psi_t = |\psi_t^0| \exp\left[4\mathrm{i}(\theta_a - \mu_a t)\right], \\ \psi_p = |\psi_p^0| \exp\left[5\mathrm{i}(\theta_a - \mu_a t)\right], \end{cases} \tag{5.12}$$

其中, μ_a 是原子的化学势. 将方程组 (5.12) 代入方程组 (5.11), 同时保持中间态无布居数分布, 就可以获得下述 CPT 态解:

5.2 超冷 N 体 Efimov 多聚物分子的形成: 广义受激拉曼绝热通道技术的量子操控

$$\begin{cases} |\psi_a^0|^2 = \dfrac{-2\times 15^{\frac{1}{3}} + 2^{\frac{1}{3}}\left[45\dfrac{\lambda}{\Omega} + \sqrt{60+\left(45\dfrac{\lambda}{\Omega}\right)^2}\right]^{\frac{2}{3}}}{30^{\frac{2}{3}}\dfrac{\lambda}{\Omega}\left[45\dfrac{\lambda}{\Omega} + \sqrt{60+\left(45\dfrac{\lambda}{\Omega}\right)^2}\right]^{\frac{1}{3}}}, \\ |\psi_t^0|^2 = 0, \\ |\psi_p^0|^2 = \dfrac{1}{5}(1-|\psi_a^0|^2), \end{cases} \tag{5.13}$$

其中, 总粒子数守恒, $|\psi_a|^2 + 4|\psi_t|^2 + 5|\psi_p|^2 = 1$. 化学势和广义的双光子共振条件[42]分别为

$$\mu_a = -2\chi_{aa}|\psi_a^0|^2 - 2\chi_{ap}|\psi_p^0|^2, \tag{5.14}$$

和

$$\Delta = -\delta + (10\chi_{aa}-2\chi_{pa})|\psi_a^0|^2 + (10\chi_{ap}-2\chi_{pp})|\psi_p^0|^2. \tag{5.15}$$

满足方程 (5.13) 的解有一个非凡的性质, 只要双光子共振条件 (5.15) 能被动力学操控, 当 λ/Ω 由 0 到 ∞ 变化时, 分别对应着全部的原子态到全部五聚物分子态的变化.

现在开始考虑异核五聚物的情形. 相似地, 描述系统动力学的哈密顿量为

$$\begin{aligned}\hat{H} = -\hbar \int \mathrm{d}\boldsymbol{r} \Big\{ &\sum_{i,j}\chi'_{i,j}\hat{\psi}_i^\dagger(\boldsymbol{r})\hat{\psi}_j^\dagger(\boldsymbol{r})\hat{\psi}_j(\boldsymbol{r})\hat{\psi}_i(\boldsymbol{r}) + \delta\hat{\psi}_t^\dagger(\boldsymbol{r})\hat{\psi}_t(\boldsymbol{r}) \\ &+ \lambda'[\hat{\psi}_t^\dagger(\boldsymbol{r})\hat{\psi}_a(\boldsymbol{r})\hat{\psi}_a(\boldsymbol{r})\hat{\psi}_a(\boldsymbol{r}) + \text{H.c.}] \\ &+ (\Delta+\delta)\hat{\psi}_p^\dagger(\boldsymbol{r})\hat{\psi}_p(\boldsymbol{r}) - \Omega'[\hat{\psi}_p^\dagger(\boldsymbol{r})\hat{\psi}_t(\boldsymbol{r})\hat{\psi}_b(\boldsymbol{r}) + \text{H.c.}]\Big\}, \end{aligned} \tag{5.16}$$

其中, 下标 b 代表原子 B. 平均场近似下, 系统的运动方程变为

$$\begin{cases} \dfrac{\mathrm{d}\psi_a}{\mathrm{d}t} = 2\mathrm{i}\sum_j \chi_{aj}|\psi_j|^2\psi_a + 4\mathrm{i}\lambda\psi_t\psi_a^{*3}, \\ \dfrac{\mathrm{d}\psi_b}{\mathrm{d}t} = 2\mathrm{i}\sum_j \chi_{bj}|\psi_j|^2\psi_b - \mathrm{i}\Omega\psi_p\psi_t^*, \\ \dfrac{\mathrm{d}\psi_t}{\mathrm{d}t} = 2\mathrm{i}\sum_j \chi_{tj}|\psi_j|^2\psi_t + (\mathrm{i}\delta-\gamma)\psi_t + \mathrm{i}\lambda\psi_a^4 - \mathrm{i}\Omega\psi_p\psi_b^*, \\ \dfrac{\mathrm{d}\psi_p}{\mathrm{d}t} = 2\mathrm{i}\sum_j \chi_{pj}|\psi_j|^2\psi_p + \mathrm{i}(\Delta+\delta)\psi_p - \mathrm{i}\Omega\psi_t\psi_b, \end{cases} \tag{5.17}$$

其中，$\chi_{ij} = n\chi'_{ij}$；$\lambda = n\sqrt{n}\lambda'$；$\Omega = \sqrt{n}\Omega'$ 是重新标准化的量，衰变率 γ 代表未俘获的四聚物态上粒子的损失.

通过稳态变换：

$$\begin{cases} \psi_a = |\psi_a^0| \exp\left[i(\theta_a - \mu_a t)\right], \\ \psi_b = |\psi_b^0| \exp\left[i(\theta_b - \mu_b t)\right], \\ \psi_t = |\psi_t^0| \exp\left[4i(\theta_a - \mu_a t)\right], \\ \psi_p = |\psi_p^0| \exp\left[i\left(4\theta_a + \theta_b - (4\mu_a + \mu_b)t\right)\right], \end{cases} \tag{5.18}$$

其中，μ_b 是原子 B 的化学势. 人们容易获得下述 CPT 解：

$$\begin{cases} |\psi_b^0|^2 = \dfrac{-5 \times 3^{\frac{1}{3}} + 5^{\frac{1}{3}}\left[72\dfrac{\lambda}{\Omega} + \sqrt{75 + \left(72\dfrac{\lambda}{\Omega}\right)^2}\right]^{\frac{2}{3}}}{16 \times 15^{\frac{2}{3}} \dfrac{\lambda}{\Omega}\left[72\dfrac{\lambda}{\Omega} + \sqrt{75 + \left(72\dfrac{\lambda}{\Omega}\right)^2}\right]^{\frac{1}{3}}}, \\ |\psi_a^0|^2 = 4|\psi_b^0|^2, \\ |\psi_t^0|^2 = 0, \\ |\psi_p^0|^2 = \dfrac{1}{5} - |\psi_b^0|^2, \end{cases} \tag{5.19}$$

其中，已经运用了粒子数守恒：$|\psi_a|^2 + |\psi_b|^2 + 4|\psi_t|^2 + 5|\psi_p|^2 = 1$. 原子的化学势为

$$\begin{cases} \mu_a = -2\chi_{aa}|\psi_a^0|^2 - 2\chi_{ab}|\psi_b^0|^2 - 2\chi_{ap}|\psi_p^0|^2, \\ \mu_b = -2\chi_{ba}|\psi_a^0|^2 - 2\chi_{bb}|\psi_b^0|^2 - 2\chi_{bp}|\psi_p^0|^2. \end{cases} \tag{5.20}$$

广义的双光子共振条件为

$$\begin{aligned} \Delta = &-\delta + (8\chi_{aa} + 2\chi_{ba} - 2\chi_{pa})|\psi_a^0|^2 \\ &+ (8\chi_{ab} + 2\chi_{bb} - 2\chi_{pb})|\psi_b^0|^2 \\ &+ (8\chi_{ap} + 2\chi_{bp} - 2\chi_{pp})|\psi_p^0|^2. \end{aligned} \tag{5.21}$$

相似地，从方程 (5.19) 和方程 (5.21) 可以发现，通过动力学操控双光子共振条件，当 $\lambda/\Omega \to 0$ 和 $\lambda/\Omega \to \infty$ 时，分别对应着原子态或五聚物分子态.

2. CPT 态的线性不稳定性和绝热保真度

存在 CPT 态，却不能保证这个态能绝热跟随. 本节将研究原子–五聚物 CPT 态的线性稳定性和绝热保真度.

5.2 超冷 N 体 Efimov 多聚物分子的形成: 广义受激拉曼绝热通道技术的量子操控

采用线性稳定性分析理论, 即在稳态 CPT 解基础上增加一个小的涨落, 让系统动力学演化, 考察涨落是否对系统无影响. 为此, 线性化包含化学势的运动方程, 获得原子-五聚物转化系统在不动点 (CPT 态) 周围的 Jacobi 矩阵. 线性化方程激发频率 (对应于 Jacobi 矩阵的特征值) 的非零频模就能够解析地获得 (其中零频模对应于 Goldstone 模 [85,86]:

$$\omega = \pm\sqrt{\frac{B \pm \sqrt{B^2 - 4C}}{2}}, \tag{5.22}$$

其中,

$$B = \begin{cases} 32\lambda^2|\psi_a^0|^6 + 2(|\psi_a^0|^2 - |\psi_p^0|^2)\Omega^2 + A^2, \text{同核系统}, \\ 32\lambda^2|\psi_a^0|^6 + 2(|\psi_b^0|^2 - |\psi_p^0|^2)\Omega^2 + A^2, \text{异核系统}, \end{cases} \tag{5.23}$$

$$C = \begin{cases} \begin{aligned}&\left[16\lambda^2|\psi_a^0|^6 + (|\psi_a^0|^2 - |\psi_p^0|^2)\Omega^2\right]^2 \\ &- 4A[8\lambda\Omega(\chi_{aa} - \chi_{ap})|\psi_a^0|^5|\psi_p^0| \\ &+ \Omega^2(\chi_{aa} - 2\chi_{ap} + \chi_{pp})|\psi_a^0|^2|\psi_p^0|^2 + 16\lambda^2\chi_{aa}|\psi_a^0|^8],\end{aligned} \\ \text{同核系统}, \\ \begin{aligned}&\left[-16\lambda^2|\psi_a^0|^6 + (|\psi_b^0|^2 - |\psi_p^0|^2)\Omega^2\right]^2 \\ &- 4A[8\lambda\Omega(\chi_{ab} - \chi_{ap})|\psi_a^0|^4|\psi_b^0||\psi_p^0| \\ &+ \Omega^2(\chi_{bb} - \chi_{bp} + \chi_{pp})|\psi_b^0|^2|\psi_p^0|^2 + 16\lambda^2\chi_{aa}|\psi_a^0|^8],\end{aligned} \\ \text{异核系统}, \end{cases} \tag{5.24}$$

这里, 对同核原子-五聚物系统 $A = \delta + 2(\chi_{at} - 4\chi_{aa})|\psi_a^0|^2 + 2(\chi_{tp} - 4\chi_{ap})|\psi_p^0|^2$, 而对于异核系统 $A = \delta + 2(\chi_{at} - 4\chi_{aa})|\psi_a^0|^2 + 2(\chi_{bt} - 4\chi_{ab})|\psi_b^0|^2 + 2(\chi_{tp} - 4\chi_{ap})|\psi_p^0|^2$. 当 ω 变为复数时, 对应的 CPT 态是动力学不稳定的. 因此, 不稳定的区域要么满足 $C < 0$, 要么满足 $C > B^2/4$. 可以发现 CPT 态的稳定性强烈依赖于非线性相互作用. 一些感兴趣的参数下有关稳定性的结果总结在图 5.7 中, 可以看到 (Ω, δ) 空间将整个区域分成稳定 (白色区域) 和不稳定 (灰色区域) 区域两部分. 存在两个不稳定区域: 区域 I 沿着 Ω 轴, 比较细长, 对应由 $C > B^2/4$ 获得的不稳定区域; 区域 II 对应于由 $C < 0$ 引起的不稳定区域. 为了让超冷原子稳定地形成五聚物分子, 设计绝热通道时避免不稳定区域是至关重要的.

在计算中, 取 ^{133}Cs 和 ^{87}Rb 原子对应的有关参数. s 波散射长度对铯原子和铷原子分别为 $a = -374a_0$ [78] 和 $a = 100a_0$ [87] (a_0 是玻尔半径). 原子密度为 $n = 6 \times 10^{19}\text{m}^{-3}$, $\lambda = 1.961 \times 10^4\text{s}^{-1}$. 因此, 可以得到相互作用参数 $\chi_{aa} =$

$0.182\lambda, \chi_{bb} = 0.074\lambda$，其余的相互作用参数均取为 0.055λ①.

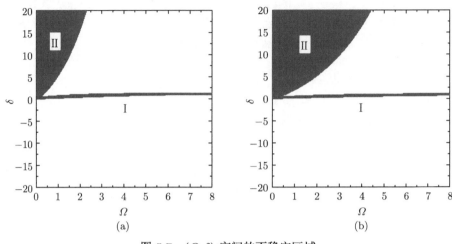

图 5.7 (Ω, δ) 空间的不稳定区域

(a) 同核系统; (b) 异核系统

直接数值求解包含损失项的方程 (5.11) 和方程 (5.17), 一些原子–五聚物分子转化的结果如图 5.8 所示, 图 5.8(a) 和 (b) 分别表示同核和异核的情况. 没有考虑粒子间两体相互作用时的保真度也分别如图 5.8 (c) 和 (d) 所示, 图中参数 $\delta = -2.0$, $\Omega_0 = 50, \tau = 20, \gamma = 1.0$. 随时间演化的 Rabi 频率为

$$\Omega(t) = \Omega_0 \text{sech} \frac{t}{\tau}, \tag{5.25}$$

其中, Ω_0 和 τ 分别是 Rabi 脉冲的强度和宽度. 选取合适的 δ 值使得系统保持在稳定区域. 值得注意的是, 时间的单位是 $1/\lambda$, 其他参数的量纲以 λ 为单位. 同核和异核五聚物系统对应的 CPT 态的解析解 (5.13) 和 (5.19) 也如图 5.8 所示. 可以发现, 通过优化系统的参数, 稳定地形成五聚物分子总是可以实现的.

在稳定区域, CPT 态的存在使得原子和五聚物分子能够发生绝热相干布居数转化. 整个系统的绝热演化可以通过引入绝热保真度 [88,89] 来研究, 绝热保真度描述了绝热解和真实解之间的距离. 这里, 定义原子–五聚物分子转化系统 CPT 态的绝热保真度为

$$F^{ap}(t) = |\langle \overline{\psi(t)} | \overline{\text{CPT}} \rangle|^2, \tag{5.26}$$

---

① 需要注意的是, 到目前为止关于多原子分子散射长度还没有好的计算办法, 于是取所有关于多原子分子相互作用参数都和原子–双原子分子相互作用的值相同. 在计算中, 其他的系数都重新归一化为以 λ 为量纲的量.

5.2 超冷 N 体 Efimov 多聚物分子的形成: 广义受激拉曼绝热通道技术的量子操控

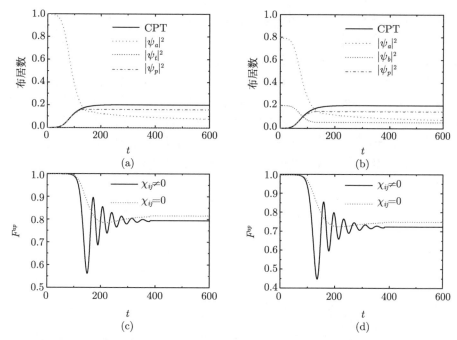

图 5.8 同核 (左图) 和异核 (右图) 系统布居数 (上图) 和绝热保真度 (下图) 随时间的演化

其中, $|\psi(t)\rangle$ 是 Schrödinger 方程的精确解; $|\overline{\psi(t)}\rangle$ 和 $|\overline{\text{CPT}}\rangle$ 是 $|\psi(t)\rangle$ 和 CPT 态重新尺度化的波函数. 系统的真实解是

$$|\overline{\psi(t)}\rangle = \begin{cases} \left(\dfrac{\psi_a^5}{|\psi_a|^4}, 2\dfrac{\psi_a\psi_t}{|\psi_a|}, \sqrt{5}\psi_p\right)^{\text{T}}, \text{同核系统}, \\ \left(\dfrac{\psi_a^4\psi_b}{|\psi_a|^3|\psi_b|}, \dfrac{\psi_a^4\psi_b}{|\psi_a|^4}, \dfrac{2\psi_b\psi_t}{|\psi_b|}, \sqrt{5}\psi_p\right)^{\text{T}}, \text{异核系统}. \end{cases} \quad (5.27)$$

如果系统能跟随 CPT 态绝热演化, 绝热保真度的值应该保持为 1. 四聚物作为中间暗态技术的同核和异核原子-五聚物分子转化系统的绝热保真度随时间的变化如图 5.8(c) 和 (d) 所示. 从图中可以看到, 在初始时间里, 系统绝热地演化, 当时间在 150 (同核系统) 和 135 (异核系统) 时, 绝热保真度分别到达最小值 0.56 和 0.45, 这意味着在此时系统远远偏离 CPT 态. 之后, 绝热保真度开始出现涨落, 其最终值分别为 0.795 和 0.723, 对应的没有考虑粒子间相互作用的结果也显示在该图上. 不难发现, 粒子间相互作用抑制了分子的转化.

进一步发现绝热保真度的最终值 $F^{ap}(\infty)$ 可以用来表示原子-五聚物分子的转化效率, 因为 $F^{ap}(\infty)$ 是绝热保真度 $F^{ap}(t)$ 在这个演化过程中的最终值, 大的 $F^{ap}(\infty)$ 值, 表示高的转化效率.

转化效率对外场参数的依赖关系如图 5.9 所示,没有考虑两体相互作用时对应的结果为图中短虚线所示. 可以看到,在红失谐 ($\delta < 0$) 时,稳定的形成超冷五聚物分子总是可能的,而在蓝失谐 ($\delta > 0$) 时,最终的转化效率却很小. 不过,不论红失谐还是蓝失谐,当粒子间两体相互作用不存在时,总有高的分子转化效率. 随着 Rabi 脉冲强度和宽度的增加,在 $\Omega_0(\Omega_0 < 10)$ 和 $\tau(\tau < 10)$ 时分子转化效率很快地增大,而在大的 Ω_0 和 τ 时,增加缓慢. 与考虑两体相互作用的情况对比,分子转化效率降低了,即两体相互作用抑制了五聚物分子的转化. 很明显,从超冷原子到五聚物分子的转化效率可以通过合理地控制外场参数来实现.

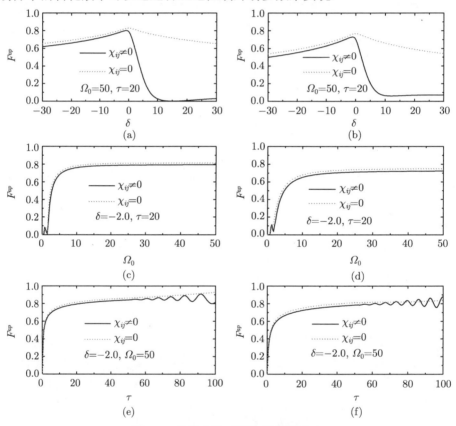

图 5.9 外场参数对转化效率的影响

同核系统 (左) 和异核系统 (右). (a), (b) 绝热保真度随失谐 δ 的变化; (c), (d) 绝热保真度随 Rabi 脉冲强度 Ω_0 的变化; (e), (f) 绝热保真度随脉冲宽度 τ 的变化

3. 讨论

应该指出的是,到目前为止还没有具体提到如何形成中间四聚物态. 近年来,在超冷铯原子气体中,实验上已经实现了和 Efimov 物理密切相关的具有统一性质

的四聚物态 [78], 这提供了形成中间四聚物态的可能的好方式. 然而, 要在实验上实现完全的原子-五聚物分子的转化还是具有挑战性, 本章中分析的一些条件也很难满足. 不过, 随着产生和操控超冷分子技术和水平的不断提高 [77,80,90,91] (包括在光晶格中形成长寿命、高效超冷分子 [92,93], 以及稳定的双原子分子、三原子分子、四聚物分子在低维几何中的发现 [94,95]), 该方法有望在将来的实验中实现.

5.2.2 超冷 N 体 Efimov 多聚物分子的形成

从前面可知, 利用广义的受激拉曼绝热通道技术, 结合 Efimov 共振, 人们已经获得了三体 Efimov 分子 [71]、四聚物分子 [83,84] 和五聚物分子 [32]. 随着实验水平和理论手段的不断改进和提高, 多体 Efimov 共振现象的研究将不断深入 [76,79-82]. 因此, 在本节中采用和 5.2.1 节相似的受激拉曼绝热通道技术, 从理论上研究超冷玻色系统中 N 体多原子-分子的转化问题. 原子首先缔合成激发态的多原子分子 A_{N-1}, 然后和另一个原子通过光场缔合成束缚的 N 体多聚物分子, 以这种方式产生的分子称之为 Efimov 分子. 首先介绍了形成同核和异核多聚物分子的物理模型, 之后得到了系统的相干布居数俘获态, 即暗态解, 发现它满足一个简单的代数方程. 进一步聚焦于原子-分子暗态的线性不稳定性和绝热保真度, 发现粒子间的相互作用导致了系统的线性不稳定性, 同时诱发了绝热保真度的振荡, 形成了一个振荡窗口. 多聚物的原子数 N 和外场参数将影响绝热保真度振荡的持续时间和振幅 [33].

1. 模型和 CPT 态

本节所考虑的模型如图 5.10 所示. 首先通过两体 Feshbach 共振 ($N = 2$) 或多体 Efimov 共振 ($N \geqslant 3$) 由超冷玻色原子耦合到多原子分子 A_{N-1}; 然后将 A_{N-1} 和另一个原子 A 或 B 利用光缔合方法形成同核多聚物分子 A_N 或异核多聚物分子 $A_{N-1}B$.

首先考虑形成同核多聚物分子的情形. 将 λ' 定义为原子-多聚物分子的耦合强度, 对应的失谐是 δ, 光缔合的激光 Rabi 频率为 Ω', 其失谐为 Δ. 在相互作用绘景下, 该系统的哈密顿量为

$$\hat{H} = -\hbar \int d\boldsymbol{r} \Big\{ \sum_{i,j} \chi'_{i,j} \hat{\psi}_i^\dagger(\boldsymbol{r}) \hat{\psi}_j^\dagger(\boldsymbol{r}) \hat{\psi}_j(\boldsymbol{r}) \hat{\psi}_i(\boldsymbol{r}) + \delta \hat{\psi}_m^\dagger(\boldsymbol{r}) \hat{\psi}_m(\boldsymbol{r}) \\
+ \lambda' [\hat{\psi}_m^\dagger(\boldsymbol{r})(\hat{\psi}_a(\boldsymbol{r}))^{N-1} + \text{H.c.}] + (\Delta + \delta) \hat{\psi}_p^\dagger(\boldsymbol{r}) \hat{\psi}_p(\boldsymbol{r}) \\
- \Omega' [\hat{\psi}_p^\dagger(\boldsymbol{r}) \hat{\psi}_m(\boldsymbol{r}) \hat{\psi}_a(\boldsymbol{r}) + \text{H.c.}] \Big\}, \tag{5.28}$$

其中, $\hat{\psi}_i$ 和 $\hat{\psi}_i^\dagger$ 是湮灭和产生算符; $\chi'_{i,j}$ 代表两体相互作用, $i,j = a,m,p$ 分别表示原子态、中间多聚物分子态 A_{N-1} 和最终的多聚物分子态 A_N.

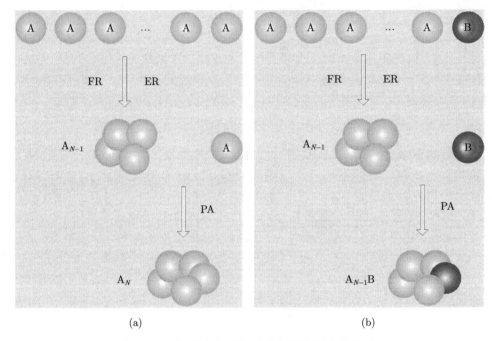

图 5.10 超冷原子–多聚物分子转化技术示意图

(a) 同核多聚物情形; (b) 异核多聚物情形

平均场近似下,场算符 $\hat{\psi}_i$ 和 $\hat{\psi}_i^\dagger$ 可以用序参数场下的 c 数 $\sqrt{n}\psi_i$ 和 $\sqrt{n}\psi_i^*$ 代替,其中,n 是总粒子数密度. 因此, 湮灭算符 $\hat{\psi}_i$ 所描述的系统运动演化的 Heisenberg 方程为

$$\begin{cases} \dfrac{\mathrm{d}\psi_a}{\mathrm{d}t} = 2\mathrm{i}\sum_j \chi_{aj}|\psi_j|^2\psi_a + (N-1)\mathrm{i}\lambda\psi_m\psi_a^{*N-2} - \mathrm{i}\Omega\psi_p\psi_m^*, \\ \dfrac{\mathrm{d}\psi_m}{\mathrm{d}t} = 2\mathrm{i}\sum_j \chi_{mj}|\psi_j|^2\psi_m + (\mathrm{i}\delta-\gamma)\psi_m + \mathrm{i}\lambda\psi_a^{N-1} - \mathrm{i}\Omega\psi_p\psi_a^*, \\ \dfrac{\mathrm{d}\psi_p}{\mathrm{d}t} = 2\mathrm{i}\sum_j \chi_{pj}|\psi_j|^2\psi_p + \mathrm{i}(\Delta+\delta)\psi_p - \mathrm{i}\Omega\psi_m\psi_a, \end{cases} \quad (5.29)$$

其中,$\chi_{ij} = n\chi'_{ij}$,$\lambda = n^{\frac{N}{2}-1}\lambda'$,$\Omega = \sqrt{n}\Omega'$,均是重新标准化的量. 引入正比于 γ 的项表示中间多聚物分子态上粒子数的损失.

为了获得具有 $|\psi_m^0| = 0$ 的稳态 CPT 解,首先忽略中间分子态上粒子数的损失,取 $\gamma = 0$,并通过下述稳态变换来获得方程组 (5.29) 的解 (μ_a 代表原子的化学势):

5.2 超冷 N 体 Efimov 多聚物分子的形成: 广义受激拉曼绝热通道技术的量子操控

$$\begin{cases} \psi_a = |\psi_a^0| \exp[i(\theta_a - \mu_a t)], \\ \psi_m = |\psi_m^0| \exp[(N-1)i(\theta_a - \mu_a t)], \\ \psi_p = |\psi_p^0| \exp[Ni(\theta_a - \mu_a t)], \end{cases} \quad (5.30)$$

将方程组 (5.30) 代入方程组 (5.29), 并让中间态布居数分布为 0, 人们可以获得下面的 CPT 解:

$$|\psi_m^0|^2 = 0, \quad (5.31)$$

$$|\psi_p^0|^2 = \frac{1}{N}(1 - |\psi_a^0|^2), \quad (5.32)$$

其中, $|\psi_a|^2$ 满足下面的代数方程:

$$|\psi_a^0|^2 + N\left(\frac{\lambda}{\Omega}\right)^2 (|\psi_a^0|^2)^{N-2} - 1 = 0, \quad (5.33)$$

总粒子数保持守恒, 即 $|\psi_a|^2 + (N-1)|\psi_m|^2 + N|\psi_p|^2 = 1$. 原子化学势为

$$\mu_a = -2(\chi_{aa} + \chi_{ap})|\psi_a^0|^2 \quad (5.34)$$

而广义双光子共振条件为

$$\Delta = -\delta + 2(N\chi_{aa} - \chi_{pa})|\psi_a^0|^2 + 2(N\chi_{ap} - \chi_{pp})|\psi_p^0|^2. \quad (5.35)$$

只要知道多聚物分子的原子个数 N, 就能容易地获得 CPT 解. 图 5.11(a) 表示在不

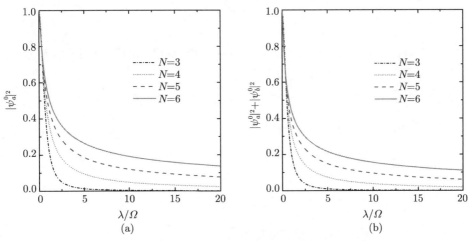

图 5.11　CPT 解中原子布居数随参数 λ/Ω 的变化

(a) 同核系统; (b) 异核系统

同的 N 时 (为了便于计算, 取 $N = 3, 4, 5, 6$ 作为例子) 满足方程 (5.33) 的原子布居数 $|\psi_a|^2$ 随参数 λ/Ω 的变化情况. 显而易见, 随着参数 λ/Ω 的增加, $|\psi_a|^2$ 减小, 直至单调地趋于零. 同时看到, N 值越小, 这种下降会更快, 这意味着不但布居数分布有一个非凡的性质, 即随着参数 λ/Ω 从 0 到 ∞ 改变, 只要双光子共振条件 (5.35) 能动力学操控, 系统将由原子态转变为多聚物分子态, 而且在相同条件下从超冷原子到更复杂的多聚物分子其转化效率要更低.

下面研究异核多聚物分子的情况, 描述该系统动力学的哈密顿量为

$$\hat{H} = -\hbar \int \mathrm{d}\boldsymbol{r} \Big\{ \sum_{i,j} \chi'_{i,j} \hat{\psi}_i^\dagger(\boldsymbol{r}) \hat{\psi}_j^\dagger(\boldsymbol{r}) \hat{\psi}_j(\boldsymbol{r}) \hat{\psi}_i(\boldsymbol{r}) + \delta \hat{\psi}_m^\dagger(\boldsymbol{r}) \hat{\psi}_m(\boldsymbol{r})$$
$$+ \lambda' \big[\hat{\psi}_m^\dagger(\boldsymbol{r}) (\hat{\psi}_a(\boldsymbol{r}))^{N-1} + \text{H.c.} \big] + (\Delta + \delta) \hat{\psi}_p^\dagger(\boldsymbol{r}) \hat{\psi}_p(\boldsymbol{r})$$
$$- \Omega' \big[\hat{\psi}_p^\dagger(\boldsymbol{r}) \hat{\psi}_m(\boldsymbol{r}) \hat{\psi}_b(\boldsymbol{r}) + \text{H.c.} \big] \Big\}, \tag{5.36}$$

其中, 下标 b 表示原子 B, $i, j = a, m, p$ 分别代表原子态、中间多聚物分子 A_{N-1} 态和多聚物分子 $A_{N-1}B$ 态. 通过运用平均场近似, 描述系统运动的方程变为

$$\begin{cases} \dfrac{\mathrm{d}\psi_a}{\mathrm{d}t} = 2\mathrm{i} \sum_j \chi_{aj} |\psi_j|^2 \psi_a + \mathrm{i}(N-1)\lambda \psi_m \psi_a^{*N-2}, \\[4pt] \dfrac{\mathrm{d}\psi_b}{\mathrm{d}t} = 2\mathrm{i} \sum_j \chi_{bj} |\psi_j|^2 \psi_b - \mathrm{i}\Omega \psi_p \psi_m^*, \\[4pt] \dfrac{\mathrm{d}\psi_m}{\mathrm{d}t} = 2\mathrm{i} \sum_j \chi_{mj} |\psi_j|^2 \psi_m + (\mathrm{i}\delta - \gamma)\psi_m + \mathrm{i}\lambda \psi_a^N - \mathrm{i}\Omega \psi_p \psi_b^*, \\[4pt] \dfrac{\mathrm{d}\psi_p}{\mathrm{d}t} = 2\mathrm{i} \sum_j \chi_{pj} |\psi_j|^2 \psi_p + \mathrm{i}(\Delta + \delta)\psi_p - \mathrm{i}\Omega \psi_m \psi_b, \end{cases} \tag{5.37}$$

其中, $\chi_{ij} = n\chi'_{ij}, \lambda = n^{\frac{N}{2}-1}\lambda', \Omega = \sqrt{n}\Omega'$ 均是重新标准化的量. 同样, γ 项表示无俘获中间多聚物态上粒子数的损失.

通过下列稳态变换:

$$\begin{cases} \psi_a = |\psi_a^0| \exp\left[\mathrm{i}(\theta_a - \mu_a t)\right], \\ \psi_b = |\psi_b^0| \exp\left[\mathrm{i}(\theta_b - \mu_b t)\right], \\ \psi_m = |\psi_m^0| \exp\left[\mathrm{i}(N-1)(\theta_a - \mu_a t)\right], \\ \psi_p = |\psi_p^0| \exp\left(\mathrm{i}\{[(N-1)\theta_a + \theta_b] - [(N-1)\mu_a + \mu_b]t\}\right), \end{cases} \tag{5.38}$$

其中, μ_b 为原子 B 的化学势. 其 CPT 解为

$$|\psi_b^0|^2 = \frac{|\psi_a^0|^2}{N-1}, \tag{5.39}$$

5.2 超冷 N 体 Efimov 多聚物分子的形成: 广义受激拉曼绝热通道技术的量子操控

$$|\psi_m^0|^2 = 0, \tag{5.40}$$

$$|\psi_p^0|^2 = \frac{1}{N} - \frac{|\psi_a^0|^2}{N-1}, \tag{5.41}$$

其中, $|\psi_a|^2$ 满足下面的代数方程:

$$\frac{N}{N-1}|\psi_a^0|^2 + N(N-1)\left(\frac{\lambda}{\Omega}\right)^2 (|\psi_a^0|^2)^{N-2} - 1 = 0. \tag{5.42}$$

这里已经采用了粒子数守恒条件: $|\psi_a|^2 + |\psi_b|^2 + (N-1)|\psi_m|^2 + N|\psi_p|^2 = 1$. 原子的化学势为

$$\begin{cases} \mu_a = -2(\chi_{aa}|\psi_a^0|^2 + \chi_{ab}|\psi_b^0|^2 + \chi_{ap}|\psi_p^0|^2), \\ \mu_b = -2(\chi_{ba}|\psi_a^0|^2 + \chi_{bb}|\psi_b^0|^2 + \chi_{bp}|\psi_p^0|^2), \end{cases} \tag{5.43}$$

广义双光子共振条件为

$$\begin{aligned} \Delta = &-\delta + 2\big[(N-1)\chi_{aa} + \chi_{ba} - \chi_{pa}\big]|\psi_a^0|^2 \\ &+ 2\big[(N-1)\chi_{ab} + \chi_{bb} - \chi_{pb}\big]|\psi_b^0|^2 \\ &+ 2\big[(N-1)\chi_{ap} + \chi_{bp} - \chi_{pp}\big]|\psi_p^0|^2. \end{aligned} \tag{5.44}$$

只要给定 N, 就能获得对应的 CPT 解.

不同 N 值时满足方程组 (5.39) 和方程 (5.42) 的原子布居数 $|\psi_a^0|^2 + |\psi_b^0|^2$ 随参数 λ/Ω 的变化情况如图 5.11(b) 所示. 类似于五聚物分子的情形, 通过动力学操控广义双光子共振条件, 布居数在极限 $\lambda/\Omega \to 0$ 和 $\lambda/\Omega \to \infty$ 对应着原子态和多聚物分子态.

为方便起见, 表 5.1 列出了 N 取不同值时, CPT 解 $|\psi_a^0|^2$ 的表达式. (由于 $N=6$ 时表达式太复杂, 没有列出).

表 5.1 CPT 解 ($|\psi_a^0|^2$)

N	同核系统	异核系统
3	$\dfrac{\Omega^2}{\Omega^2 + 3\lambda^2}$	$\dfrac{2\Omega^2}{3(4\lambda^2 + \Omega^2)}$
4	$\dfrac{-\Omega^2 + \Omega\sqrt{16\lambda^2 + \Omega^2}}{8\lambda^2}$	$\dfrac{-\Omega^2 + \Omega\sqrt{27\lambda^2 + \Omega^2}}{18\lambda^2}$
5	$\dfrac{-2 \times 15^{\frac{1}{3}} + 2^{\frac{1}{3}}\left[45\dfrac{\lambda}{\Omega} + \sqrt{60 + \left(45\dfrac{\lambda}{\Omega}\right)^2}\right]^{\frac{2}{3}}}{30^{\frac{2}{3}}\dfrac{\lambda}{\Omega}\left[45\dfrac{\lambda}{\Omega} + \sqrt{60 + \left(45\dfrac{\lambda}{\Omega}\right)^2}\right]^{\frac{1}{3}}}$	$\dfrac{-5 \times 3^{\frac{1}{3}} + 5^{\frac{1}{3}}\left[72\dfrac{\lambda}{\Omega} + \sqrt{75 + \left(72\dfrac{\lambda}{\Omega}\right)^2}\right]^{\frac{2}{3}}}{4 \times 15^{\frac{2}{3}}\dfrac{\lambda}{\Omega}\left[72\dfrac{\lambda}{\Omega} + \sqrt{75 + \left(72\dfrac{\lambda}{\Omega}\right)^2}\right]^{\frac{1}{3}}}$
...	...	...

2. CPT 态的稳定性和绝热性

同样, 存在 CPT 态并不能保证系统能绝热跟随这个态. 因此, 分析不稳定性和绝热保真度就至关重要. 本小节将讨论原子–多聚物分子 CPT 态的线性不稳定性和绝热保真度.

在广义受激拉曼绝热通道技术中, 避免不稳定区域是非常必要的. 为此, 线性化包含系统化学势的运动方程, 首先获得原子–多聚物分子转化系统在不动点 (CPT 态) 周围的 Jacobi 矩阵. 激发频率 (对应于 Jacobi 矩阵的特征值) 的非零频模就能够解析地获得 (其中零频模对应于 Goldstone 模 [86]),

$$\omega = \pm\sqrt{\frac{B \pm \sqrt{B^2 - 4C}}{2}}, \tag{5.45}$$

其中,

$$B = \begin{cases} 2N(N-2)\lambda^2|\psi_a^0|^{2(N-2)} + 2|\psi_a^0|^2\Omega^2 + A^2, & \text{同核系统}, \\ 2(N-1)(N-2)\lambda^2|\psi_a^0|^{2(N-2)} + \dfrac{2}{N-1}\Omega^2|\psi_a^0|^2 + A^2, & \text{异核系统}, \end{cases} \tag{5.46}$$

$$C = \begin{cases} \left[N(N-2)\lambda^2|\psi_a^0|^{2(N-2)} + |\psi_a^0|^2\Omega^2\right]^2 \\ \quad + 4A\lambda^2(N^2\chi_{aa} - 2N\chi_{ap} + \chi_{pp})|\psi_a^0|^{2(N-1)}, & \text{同核系统}, \\ \left[(N-1)(N-2)\lambda^2|\psi_a^0|^{2(N-2)} + \dfrac{1}{N-1}|\psi_a^0|^2\Omega^2\right]^2 \\ \quad + 4A\lambda^2|\psi_a^0|^{2(N-1)}\big[(N-1)^2\chi_{aa} + 2(N-1)\chi_{ab} \\ \quad - 2(N-1)\chi_{ap} + \chi_{bb} - 2\chi_{bp} + \chi_{pp}\big], & \text{异核系统}, \end{cases} \tag{5.47}$$

这里, 对同核系统 $A = -\delta + 2[(N-1)\chi_{aa} - \chi_{am}]|\psi_a^0|^2 + 2[(N-1)\chi_{ap} - \chi_{mp}]|\psi_p^0|^2$, 而对异核原子–多聚物分子系统 $A = -\delta + 2[(N-1)\chi_{aa} - \chi_{am}]|\psi_a^0|^2 + 2[(N-1)\chi_{ab} - \chi_{bm}]|\psi_b^0|^2 + 2[(N-1)\chi_{ap} - \chi_{mp}]|\psi_p^0|^2$.

当 ω 为复数时, 对应的 CPT 态是动力学不稳定的. 因此, 不稳定范围由 $C < 0$ 或 $C > B^2/4$ 给出. 图 5.12 所示为在一些参数下对稳定性分析的典型结果. 将 (Ω, δ) 空间划分成稳定区域和不稳定区域, 其中不稳定区域分为两部分: 范围 I 沿着 Ω 轴, 比较窄, 对应于 $C > B^2/4$, 随着 N 增加, 该区域扩展并逐渐上移; 范围 II 发生在小 Ω 的情形, 满足 $C < 0$, 随着 N 的增加, 对同核系统该区域宽度会变窄, 而对于异核系统, 该区域几乎不变. 为了转化原子到稳定的多聚物分子, 在设计绝热通道时避免不稳定区域是至关重要的.

5.2 超冷 N 体 Efimov 多聚物分子的形成: 广义受激拉曼绝热通道技术的量子操控

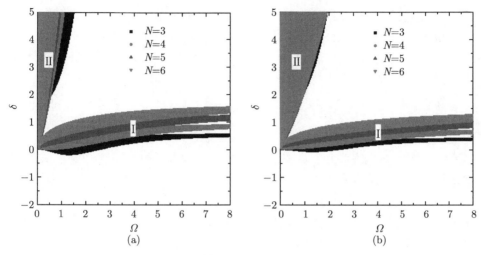

图 5.12 原子–多聚物分子系统 (Ω, δ) 空间的不稳定区域

(a) 同核系统; (b) 异核系统

和 5.2.1 节一样, 在计算中选取 ^{133}Cs 和 ^{87}Rb 原子来形成多聚物分子 Cs$_N$ 和 Cs$_{N-1}$Rb. s 波散射长度对铯原子和铷原子分别为 $a = -374a_0$ [78] 和 $a = 100a_0$ [87] (a_0 是玻尔半径). 原子密度为 $n = 6 \times 10^{19} \text{m}^{-3}$, $\lambda = 1.961 \times 10^4 \text{s}^{-1}$. 这样可以得到相互作用参数 $\chi_{aa} = 0.182\lambda$, $\chi_{bb} = 0.074\lambda$, 其余的相互作用参数均取为 0.055λ [32]. 注意到, 时间的单位是 $1/\lambda$, 而其余系数的单位为 λ.

在稳定区域, 存在 CPT 态就使得原子和多聚物分子之间可以进行绝热相干转化, 布居数随时间的演化可以通过直接求解方程组 (5.29) 和方程组 (5.37) 获得. 布居数 $|\psi_p|^2$ 的数值结果如图 5.13 所示, 图中也显示了方程 (5.32) 和方程 (5.41) 对

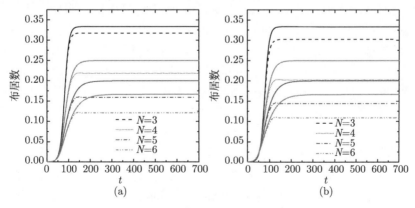

图 5.13 布居数随时间的变化

(a) 同核系统; (b) 异核系统

应的 CPT 态的解析解, 如实线所示, 其他虚线表示布居数 $|\psi_p|^2$, 其他参数分别为 $\delta = -2.0, \Omega_0 = 50, \tau = 20, \gamma = 1.0$. 从图 5.13 中可以看到, 在初始的时间里, 布居数动力学能够完全跟随预言的 CPT 解. 然而, 后来二者出现了差别, 而且随着参数 N 的增大, 这种偏离越来越大. 在计算中, Rabi 频率调制形式取为

$$\Omega(t) = \Omega_0 \mathrm{sech}\frac{t}{\tau}, \tag{5.48}$$

其中, Ω_0 是脉冲强度; τ 是脉冲宽度; δ 的取值要使得系统保持在稳定区域.

系统真实的演化态和 CPT 态之间的偏离意味着系统不能完全绝热跟随, 这是由绝热损失和中间多聚物分子态的损失造成的. 在 CPT 解中, 为方便计算, 取 $\gamma = 0$, 而在数值计算中, 取 $\gamma = 1.0$. 事实上, 系统的这种绝热演化完全能够采用绝热保真度来描述 [32,88,89], 绝热保真度就是指绝热解和真实解之间的距离. 这里, 定义原子-多聚物分子 CPT 态的绝热保真度为

$$F^{ap}(t) = |\langle \overline{\psi(t)} | \overline{\mathrm{CPT}} \rangle|^2, \tag{5.49}$$

其中, $|\psi(t)\rangle$ 是 Schrödinger 方程的精确解. $|\overline{\psi(t)}\rangle$ 和 $|\overline{\mathrm{CPT}}\rangle$ 分别是 $|\psi(t)\rangle$ 和 CPT 态重新尺度化的波函数. 系统的真实态为

$$|\overline{\psi(t)}\rangle = \begin{cases} \left(\dfrac{\psi_a^N}{|\psi_a|^{N-1}}, \sqrt{N-1}\dfrac{\psi_a\psi_m}{|\psi_a|}, \sqrt{N}\psi_p\right)^{\mathrm{T}}, & \text{同核系统}, \\ \left(\dfrac{\psi_a^{N-1}\psi_b}{|\psi_a|^{N-2}|\psi_b|}, \dfrac{\psi_a^{N-1}\psi_b}{|\psi_a|^{N-1}}, \sqrt{N-1}\dfrac{\psi_b\psi_m}{|\psi_b|}, \sqrt{N}\psi_p\right)^{\mathrm{T}}, & \text{异核系统}. \end{cases} \tag{5.50}$$

图 5.14(a) 和 (b) 分别表示多聚物暗态技术下同核和异核系统绝热保真度随时间的变化情况, 参数和图 5.13 一致. 演化过程沿着时间轴分成三个窗口. 第一个窗口位于初始时间段, 绝热保真度的值接近 1, 这意味着系统在这个区域能绝热跟随 CPT 态. 随后, 进入第二个窗口——振荡窗口, 绝热保真度首先明显地下降到一个最小值, 然后随时间振荡, 其幅度也逐渐变小, 最后绝热保真度接近一个小于 1 的稳定值, 此时对应于第三个窗口——稳定窗口. 可以发现, 整个演化过程强烈地依赖于原子数 N、粒子间相互作用和外场参数. 随着 N 的增加, 不论同核系统还是异核系统, 振荡窗口都将加宽, 振荡幅度也将增加, 而最终的绝热保真度的值会下降. 这说明, 在相同条件下, 从超冷原子形成更复杂的超冷分子, 其转化效率要更低些. 进一步研究表明: ① 绝热保真度的振荡是由粒子间相互作用引起的; ② 随着脉冲宽度 τ 的增加, 第一和第二个窗口将展宽 (振荡窗口的宽度与 τ 成正比), 并且振荡幅度将降低, 而最终的绝热保真度的值将增加, 这说明只要合理优化系统参数, 稳定形成多聚物分子总是可以实现的. 以同核五聚物为例, 数值计算了绝热保真度随参数 τ 的变化情况, 也计算了对应的无粒子间相互作用时绝热保真度随时间的演化, 这些结果如图 5.15(a) 和 (b) 所示, 图中取 $N = 5, \Omega_0 = 50$.

5.2 超冷 N 体 Efimov 多聚物分子的形成: 广义受激拉曼绝热通道技术的量子操控

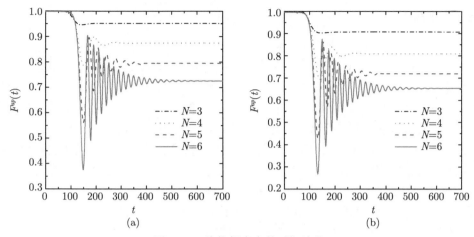

图 5.14 绝热保真度的时间演化

(a) 同核系统; (b) 异核系统

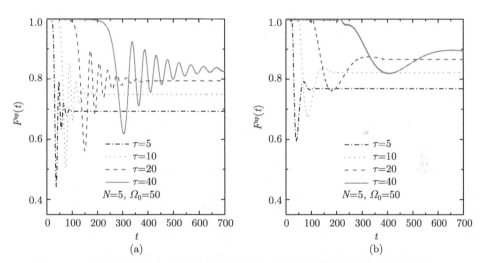

图 5.15 不同脉冲宽度 τ 时原子-五聚物分子转化系统绝热保真度的时间演化

(a) 考虑粒子间相互作用情形; (b) 没有考虑粒子间相互作用的情形

进一步研究也表明,转化效率随外场参数的变化情况和五聚物分子的变化情况相似[32]. 在红失谐 ($\delta < 0$) 时,稳定地形成多聚物分子总是可能的,而蓝失谐 ($\delta > 0$) 时,转化效率很低. 然而,不论红失谐还是蓝失谐,粒子间相互作用总是抑制转化效率. 很明显,从原子到超冷 N 体多聚物分子的转化效率可通过外场参数 δ、Ω_0 和 τ 来调控.

5.2.3 小结

总之, 本节采用广义受激拉曼绝热通道技术研究了包含两体相互作用的超冷玻色系统中同核和异核 N 体多聚物分子的转化问题. 首先研究了超冷五聚物分子的形成, 接着将这一方法进行推广, 实现了超冷 N 体多聚物分子的转化, 获得了系统的 CPT 解; 对 N 体多聚物, 获得了暗态所满足的统一的代数方程. 换句话说, 只要给出多聚物分子的原子个数 N, 就可以通过解代数方程获得 CPT 解. 通过线性稳定性理论, 分析了原子-多聚物暗态的线性不稳定性, 并通过适当的绝热保真度的定义, 研究了系统的绝热保真度, 同时也讨论了单光子失谐、Rabi 脉冲强度和宽度、粒子间相互作用以及原子个数等参数对整个转化过程的影响. 研究表明, 粒子间相互作用导致了系统的不稳定性, 引起绝热保真度的振荡, 并抑制了多聚物分子的转化效率. 多聚物分子包含的原子数将影响不稳定的范围和绝热保真度的振荡幅度. Rabi 脉冲的宽度 τ 会影响绝热保真度的振荡范围和幅度. 合理优化外场参数, 总是可以实现原子到多聚物分子的稳定转化. 研究结果不仅包含了先前产生超冷三原子分子、四聚物分子以及五聚物分子的方法, 而且更为将来在实验上产生超冷 N 体多聚物分子提供了一条可行路径.

5.3 外场形式对超冷原子-多聚物分子转化效率的影响

本节主要讨论外场形式对超冷原子-多聚物分子转化效率的影响, 从而获得稳定、高效形成超冷多原子分子的方法[96].

前面已讨论, 超冷多聚物分子的形成在研究超冷化学和超化学[52]、高精密分子谱[97]、分子物质波[47] 等方面有很强的理论价值, 并且在量子计算、原子钟等领域有着广泛的应用前景, 因此超冷原子-多聚物分子的转化一直是冷原子分子物理领域备受关注的研究课题[7]. 人们借助广义的 STIRAP 技术[42,64,98], 获得了较为有效的超冷原子-分子转化方案[99,100]. STIRAP 技术的成功依赖于相干布居俘获态[100-102] (CPT 态, 即暗态) 的存在, 在暗态上的绝热演化能有效地抑制激发态上粒子数的自发辐射, 使得原子能以很高的转化效率形成多聚物分子. 目前超冷原子-分子的转化已从起初的双原子分子[3,18,24,88,103]、三原子分子聚合[104] 拓展到 N 体多原子分子的聚合[32,33], 在先前研究所给出的外场形式下, 可以稳定地产生超冷双原子、三原子分子, 且转化效率较高. 然而在实现五聚物或 N 聚物分子时, 转化效率还不是很理想, 绝热过程中存在明显的振荡区间[32,33]. 为了减小绝热过程中出现的不稳定区域, 提高最终的转化效率, 本以五聚物分子为例, 首先将文献 [32] 和 [33] 所给出的外场引入指数项, 讨论其对转化效率的影响, 接着选取一种更优化外场形式, 很大程度上减小了振荡的宽度且使绝热保真度更接近 1, 从而提高了超冷原子–多聚物分子的转化效率以及转化过程的稳定性 [96].

5.3.1 模型与绝热保真度

1. 同核系统与异核系统

原子首先通过 Efimov 共振 [77,78] 缔合成处于激发态的多原子分子 A_{N-1}；然后将 A_{N-1} 和另一个原子 A 或 B 通过光缔合形成同核多聚物分子 A_N 或异核多聚物分子 $A_{N-1}B$. 考虑以上两种情况,定义原子与多聚物分子之间的耦合强度为 λ',相应的失谐量为 δ,光缔合的激光 Rabi 频率为 Ω',相应的失谐量为 Δ,则在相互作用绘景下,系统的哈密顿量可表示为

$$\hat{H} = -\hbar \int d\bm{r} \Big\{ \sum_{i,j} \chi'_{i,j} \hat{\psi}_i^\dagger(\bm{r}) \hat{\psi}_j^\dagger(\bm{r}) \hat{\psi}_j(\bm{r}) \hat{\psi}_i(\bm{r}) + \delta \hat{\psi}_m^\dagger(\bm{r}) \hat{\psi}_m(\bm{r}) \\
+ \lambda'[\hat{\psi}_m^\dagger(\bm{r})(\hat{\psi}_a(\bm{r}))^{(N-1)} + \text{H.c.}] \\
+ (\Delta + \delta) \hat{\psi}_p^\dagger(\bm{r}) \hat{\psi}_p(\bm{r}) - \Omega'[\hat{\psi}_p^\dagger(\bm{r}) \hat{\psi}_m(\bm{r}) \hat{\psi}_k(\bm{r}) + \text{H.c.}] \Big\} \tag{5.51}$$

其中,$\hat{\psi}_i(\bm{r})$ 和 $\hat{\psi}_i^\dagger(\bm{r})$ 分别为湮灭和产生算符；$\chi'_{i,j}$ 为各态的相互作用,下标 i, j 分别代表 k 态、m 态、p 态,即原子态、聚合过程中间态和目标态；在同核系统中 $\hat{\psi}_k(\bm{r}) = \hat{\psi}_a(\bm{r})$,异核系统中 $\hat{\psi}_k(\bm{r}) = \hat{\psi}_b(\bm{r})$. 平均场近似下,用序参数场中的 c-数 $\sqrt{n}\hat{\psi}_i(\bm{r})$ 和 $\sqrt{n}\hat{\psi}_i^*(\bm{r})$ 替换原先的场算符 $\hat{\psi}_i(\bm{r})$ 和 $\hat{\psi}_i^\dagger(\bm{r})$ (其中 n 为总粒子数密度),即可得到一组描述系统运动演化的方程,在同核系统中,原子态的波函数满足:

$$\frac{d\psi_a}{dt} = 2i \sum_j \chi_{aj} |\psi_j|^2 \psi_a + (N-1)i\lambda \psi_m \psi_a^{*N-2} - i\Omega \psi_p \psi_m^*, \tag{5.52}$$

在异核系统中,原子态的波函数满足:

$$\begin{cases} \dfrac{d\psi_a}{dt} = 2i \sum_j \chi_{aj} |\psi_j|^2 \psi_a + (N-1)i\lambda \psi_m \psi_a^{*N-2}, \\ \dfrac{d\psi_b}{dt} = 2i \sum_j \chi_{bj} |\psi_j|^2 \psi_b - i\Omega \psi_p \psi_m^*, \end{cases} \tag{5.53}$$

两种系统中,中间态和目标态的波函数均满足:

$$\begin{cases} \dfrac{d\psi_m}{dt} = 2i \sum_j \chi_{mj} |\psi_j|^2 \psi_m + (i\delta - \gamma)\psi_m + i\lambda \psi_a^{N-1} - i\Omega \psi_p \psi_k^*, \\ \dfrac{d\psi_p}{dt} = 2i \sum_j \chi_{pj} |\psi_j|^2 \psi_p + i(\Delta + \delta)\psi_p - i\Omega \psi_m \psi_k, \end{cases} \tag{5.54}$$

其中,正比于 γ 的项是用来模拟中间态上的粒子损失,$\chi_{ij} = n\chi'_{ij}$,$\lambda = n^{\frac{N}{2}-1}\lambda'$,$\Omega = \sqrt{n}\Omega'$. 为了得到 CPT 态形式的定态解,令中间态上无分子俘获,即 $|\psi_m^0| = 0$.

忽略中间分子态上粒子数的损失,取 $\gamma = 0$. 各态满足新的归一化条件: 在同核系统中 $|\psi_a^0|^2 + (N-1)|\psi_m^0|^2 + N|\psi_p^0|^2 = 1$, 而在异核系统中 $|\psi_a^0|^2 + |\psi_b^0|^2 + (N-1)|\psi_m^0|^2 + N|\psi_p^0|^2 = 1$. 通过合理的稳态变换,并引入原子的化学势 μ_a 和 μ_b, 求解薛定谔方程可以得到一组 CPT 解所满足的代数方程, 求解该方程即可得到系统的定态解、相应的原子化学势和双光子共振条件 [32,33].

2. CPT 态的绝热保真度

为了得到较高的转化效率, 希望整个演化过程中 CPT 态上的真实演化态能够在每个瞬时均与理论的 CPT 态绝热演化保持一致, 因此人们引入了绝热保真度的概念 [88,89]. 然而, 因为超冷原子-多聚物分子转化系统中系统的哈密顿量不再具有 $U(1)$ 不变性, 所以不能用传统的保真度概念来进行描述, 需要重新定义绝热保真度的概念 [32,33]. 由于系统的哈密顿量满足 $U(\phi)$ 不变性, 故保真度的形式可定义为 $F^{ap}(t) = |\langle\overline{\psi(t)}|\overline{\psi_{\text{CPT}}(t)}\rangle|^2$, 其中 $|\psi(t)\rangle$ 是系统薛定谔方程的精确解, 而 $|\overline{\psi(t)}\rangle$ 和 $|\overline{\psi_{\text{CPT}}(t)}\rangle$ 表示 $|\psi(t)\rangle$ 和 CPT 态重新尺度化的波函数, 它们分别为

$$|\overline{\psi(t)}\rangle = \left(\frac{\psi_a^N}{|\psi_a|^{N-1}}, \sqrt{N-1}\frac{\psi_a\psi_m}{|\psi_a|}, \sqrt{N}\psi_p\right)^{\text{T}} \quad (\text{同核系统}), \tag{5.55}$$

$$|\overline{\psi(t)}\rangle = \left(\frac{\psi_a^{N-1}\psi_b}{|\psi_a|^{N-2}|\psi_b|}, \frac{\psi_a^{N-1}\psi_b}{|\psi_a|^{N-1}}, \sqrt{N-1}\frac{\psi_b\psi_m}{|\psi_b|}, \sqrt{N}\psi_p\right)^{\text{T}} \quad (\text{异核系统}), \tag{5.56}$$

可以利用绝热保真度来研究整个系统的绝热演化, 如果系统能绝热地跟随 CPT 态演化, 绝热保真度的值应该保持为 1. 而演化结束时保真度的值定义为最终绝热保真度 F_f^{ap}, 它可以用来描述超冷多聚物分子的转化效率, F_f^{ap} 越接近 1, 系统的转化效率越高.

5.3.2 外场扫描形式对转化效率的影响

以五聚物分子为例, 通过改变外场形式, 来探索其对多聚物分子转化效率的影响 [96].

1. 原有外场形式的改进及讨论

在文献 [32] 和 [33] 中规定 λ 为不变量, 而 Rabi 频率取为 $\Omega = \Omega_0 \text{sech}\frac{t}{\tau}$, 其中, Ω_0 和 τ 分别指脉冲强度和脉冲宽度, 均取为常数. 对这种 Rabi 频率随时间的变化形式作出如下改进:

$$\Omega = \Omega_0 \text{sech}\left(\frac{t}{\tau}\right)^{\alpha}, \tag{5.57}$$

其中, α 定义为时间指数.

在数值计算中, 外场参数取 $\Omega_0 = 50$, $\tau = 20$, $\delta = -2$, 粒子间相互作用 $\chi_{aa} =$

5.3 外场形式对超冷原子–多聚物分子转化效率的影响

0.182λ, $\chi_{bb} = 0.074\lambda$, 取 $\dfrac{1}{\lambda}$ 为时间的单位, 其余各量均以 λ 为单位[32]. 当时间指数 α 小于 1 时, 系统的绝热保真度 $F^{ap}(t)$ 和系统误差 (即绝热保真度与理想值 1 的差值)$1 - F^{ap}(t)$ 随时间的变化关系如图 5.16 所示.

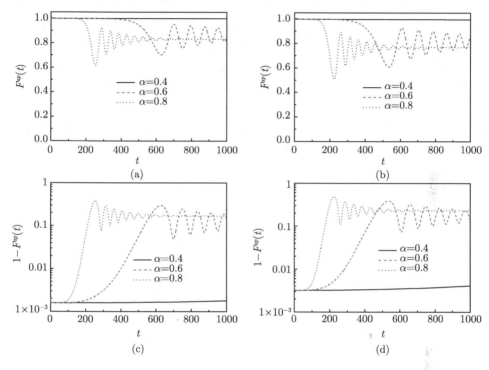

图 5.16 $F^{ap}(t)$, $1 - F^{ap}(t)$ 随时间的变化
(a)、(c) 同核系统; (b)、(d) 异核系统

可以看出, 不论同核系统还是异核系统, 在时间指数的取值较小时, 演化过程在所取的时间内只出现了一个窗口, 该窗口中绝热保真度非常接近于 1, 系统的误差较小. 当 $\alpha = 0.6$ 时, $F^{ap}(t)$ 和 $1 - F^{ap}(t)$ 随时间变化的图像在该段时间内均出现了第二个窗口——振荡窗口, 系统在该窗口的绝热过程变得不稳定, 绝热保真度在降到最小值之后开始随时间振荡, 振荡幅度也随时间逐渐减小. 当时间指数 $\alpha \geqslant 1$ 时, 系统的绝热保真度 $F^{ap}(t)$ 和与理想值的误差 $1 - F^{ap}(t)$ 随时间的变化关系如图 5.17 所示. 可以看到绝热保真度在振荡区间出现的最小值变小, 绝热保真度的振荡幅度随着时间减小并渐渐趋于稳定, 系统的误差变化也趋于一个定值, 转化过程趋于相对稳定的状态. 系统的绝热保真度在所考察的时间段内出现了第三个窗口——稳定窗口, 且随着 α 取值的进一步增大, 绝热保真度的振荡窗口变小, 能更早地到达稳定窗口. 与此同时, 系统趋于稳定后的最终绝热保真度的值也在减小,

系统误差较大,故随着 α 取值增大,系统能更快达到稳定,但最终绝热保真度变低,从而使系统的转化效率有所降低. 值得注意的是,完整的绝热演化过程中,三个窗口是不可避免的,但根据时间指数的选取,三个窗口的宽度会发生改变. 进一步研究发现,该外场中时间指数的变化对绝热保真度有很大的影响. 时间指数 α 的取值小于 1 时,对绝热保真度的振荡起到了有效的延缓作用,其值越小,振荡窗口出现得越晚. 绝热保真度在起初一段时间里保持与理想值 1 的高度接近,但随着时间的增加,振荡仍然会出现,且振荡窗口的宽度变大. 随着 α 的减小,系统稳定后的最终绝热保真度增大,系统能更加高效地进行超冷原子-多聚物分子转化,但需要较长的时间达到稳定. 而当 α 取值大于 1 时,随着 α 的增加,振荡窗口的宽度变小,绝热保真度更早地到达稳定窗口,但是系统的最终绝热保真度也同时会随着时间指数 α 的增大而变小,从而导致转化效率的降低.

图 5.18 给出了不同时间指数下目标态布居数随时间的变化情况,进一步验证了以上结论.

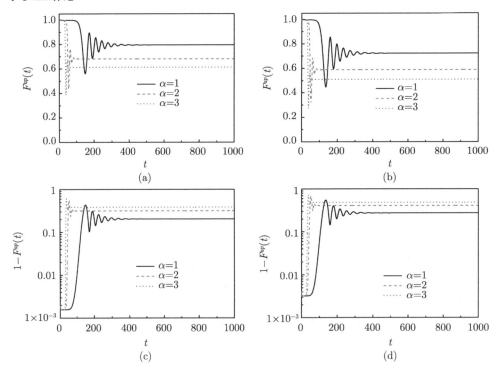

图 5.17 $F^{ap}(t)$, $1-F^{ap}(t)$ 随时间的变化

(a)、(c) 同核系统; (b)、(d) 异核系统

5.3 外场形式对超冷原子-多聚物分子转化效率的影响

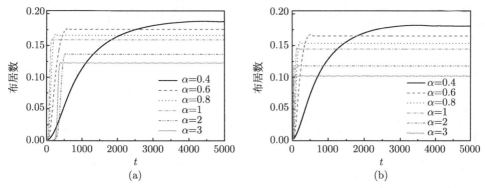

图 5.18 不同时间指数取值下的目标态布居数 $|\psi_p|^2$ 随时间的变化

(a) 同核系统; (b) 异核系统

2. 更优化的外场形式

这里选取一种更优化的外场形式, 规定 Ω 不变, 而原子-多聚物分子耦合强度 λ 的形式为

$$\lambda = \lambda_0 \cosh\frac{t}{\tau}. \tag{5.58}$$

人们已经发现该外场能高效地转化超冷三原子分子[104], 在这种外场形式下, 时间以 λ_0 为单位. 目标态布居数与 CPT 态的解析解随时间的变化如图 5.19 所示, 数值计算中取 $\Omega = 50, \tau = 20$. 通过布居数 $|\psi_p|^2$ 和 CPT 态的解析解随时间的变化图像可以看出, 布居数动力学和预言的 CPT 态解析解差别很小, 系统在观察时间内已经到达稳定窗口. 绝热保真度 $F^{ap}(t)$ 和系统误差 $1 - F^{ap}(t)$ 随时间变化的图像如图 5.20 所示.

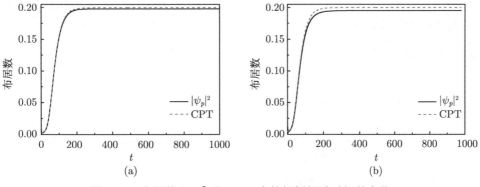

图 5.19 布居数 $|\psi_p|^2$ 和 CPT 态的解析解随时间的变化

(a) 同核系统; (b) 异核系统

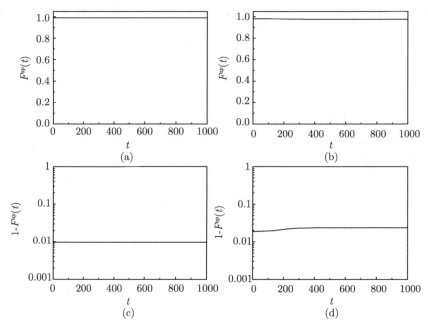

图 5.20 $F^{ap}(t)$, $1 - F^{ap}(t)$ 随时间的变化图像

(a)、(c) 同核系统; (b)、(d) 异核系统

在这种新的外场扫描形式下, 绝热保真度始终稳定在一个非常接近于 1 的数值, 系统的误差较小. 整个转化过程几乎不存在振荡窗口, 从而保证该过程系统能较理想地跟随 CPT 态绝热演化, 实现很高的转化效率. 与之前的外场扫描形式相比, 该外场既能长时间保证转化过程的稳定, 又能得到非常理想的转化效率, 从而可以实现稳定且高效的超冷原子–多聚物分子转化.

为讨论该外场形式转化效率的参数鲁棒性, 定义 $\Delta\tau_0 = \tau - \tau_0$, 数值计算了系统的最终绝热保真度 F_{f}^{ap} 随 $\Delta\tau_0/\tau_0$ 的变化情况, 如图 5.21 所示, 计算中取 $\tau_0 = 20$.

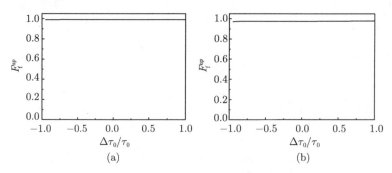

图 5.21 F_{f}^{ap} 随 $\Delta\tau_0/\tau_0$ 的变化图像

(a) 同核系统; (b) 异核系统

不难发现，随着 τ 取值的变化，系统的最终绝热保真度几乎保持不变，所以该外场参数对系统的转化效率影响较小，具有很好的参数鲁棒性．进一步研究同样发现，对多聚物分子的情况，该外场同样能稳定、高效地实现分子的转化．

5.3.3 小结

将前几节所给出的外场引入指数项并定义为时间指数，发现时间指数的变化会引起振荡窗口的宽度和最终绝热保真度的变化，从而影响系统的绝热稳定性和转化效率．当时间指数取小于 1 的值时，随着 α 取值的减小，振荡窗口变大、振荡过程中绝热保真度的最小值变大，但振荡窗口出现的时间被推迟，最终绝热保真度接近于 1．而当时间指数取大于 1 的值时，随着 α 取值的增大，振荡窗口的宽度缩小、振荡过程中绝热保真度的最小值变小，绝热保真度较早地达到稳定窗口，但是稳定后的最终绝热保真度变小，时间指数越大，最终绝热保真度越低，系统虽能更早达到稳定，但转化效率较低．另外，选取了一种更优化的外场形式，在这种外场形式下，绝热保真度几乎趋于稳定，而且保持在一个非常接近于理想值 1 的值，系统误差较小．该外场形式的参数选取对转化效率影响较小，具有很好的参数鲁棒性，在这种外场形式下可以实现更为稳定、高效的超冷原子–多聚物分子转化[96]．该研究不仅为产生超冷多聚物分子提供了理论依据，也进一步丰富了 Efimov 物理在超冷原子系统中的应用[105]．

参 考 文 献

[1] CARR L D, DEMILLE D, KREMS R V, et al. Cold and ultracold molecules: Science, technology and applications[J]. New J. Phys., 2009, 11(5): 055049.

[2] OSPELKAUS S, NI K K, WANG D, et al. Quantum-state controlled chemical reactions of ultracold potassium-rubidium molecules[J]. Science, 2010, 327(5967): 853-857.

[3] FU L B, LIU J. Adiabatic Berry phase in an atom–molecule conversion system[J]. Ann. Phys., 2010, 325(11): 2425-2434.

[4] ENOMOTO K, KASA K, KITAGAWA M, et al. Optical Feshbach resonance using the intercombination transition[J]. Phys. Rev. Lett., 2008, 101: 203201.

[5] PELLEGRINI P, GACESA M, CÔTÉ R. Giant formation rates of ultracold molecules via Feshbach-optimized photoassociation[J]. Phys. Rev. Lett., 2008, 101: 053201.

[6] MACKIE M, FENTY M, SAVAGE D, et al. Cross-molecular coupling in combined photoassociation and Feshbach resonances[J]. Phys. Rev. Lett., 2008, 101: 040401.

[7] CHIN C, GRIMM R, JULIENNE P, et al. Feshbach resonances in ultracold gases[J]. Rev. Mod. Phys., 2010, 82: 1225-1286.

[8] DONLEY E A, CLAUSSEN N R, THOMPSON S T, et al. Atom-molecule coherence in a Bose-Einstein condensate[J]. Nature, 2002, 417(6888): 529-533.

[9] THOMPSON S T, HODBY E, WIEMAN C E. Ultracold molecule production via a resonant oscillating magnetic field[J]. Phys. Rev. Lett., 2005, 95: 190404.

[10] LIU B, FU L B, LIU J. Shapiro-like resonance in ultracold molecule production via an oscillating magnetic field[J]. Phys. Rev. A, 2010, 81(1): 013602.

[11] JOCHIM S, BARTENSTEIN M, ALTMEYER A, et al. Bose-Einstein condensation of molecules[J]. Science, 2003, 302(5653): 2101-2103.

[12] XU K, MUKAIYAMA T, ABO-SHAEER J R, et al. Formation of quantum-degenerate sodium molecules[J]. Phys. Rev. Lett., 2003, 91: 210402.

[13] HODBY E, THOMPSON S T, REGAL C A, et al. Production efficiency of ultracold Feshbach molecules in Bosonic and Fermionic systems[J]. Phys. Rev. Lett., 2005, 94: 120402.

[14] MARK M, KRAEMER T, HERBIG J, et al. Efficient creation of molecules from a cesium Bose-Einstein condensate[J]. Europhys. Lett, 2005, 69(5): 706-712.

[15] ZIRBEL J J, NI K K, OSPELKAUS S, et al. Collisional stability of Fermionic Feshbach molecules[J]. Phys. Rev. Lett., 2008, 100: 143201.

[16] OLSEN M L, PERREAULT J D, CUMBY T D, et al. Coherent atom-molecule oscillations in a Bose-Fermi mixture[J]. Phys. Rev. A, 2009, 80: 030701.

[17] XU X Q, LU L H, LI Y Q. Enhancing molecular conversion efficiency by a magnetic field pulse sequence[J]. Phys. Rev. A, 2009, 80: 033621.

[18] DOU F Q, LI S C. Achieving high molecular conversion efficiency via a magnetic field pulse train[J]. Eur. Phys. J. B-Condensed Matter and Complex Systems, 2012, 85(6): 1-6.

[19] LI J, YE D F, MA C, et al. Role of particle interactions in a many-body model of Feshbach-molecule formation in bosonic systems[J]. Phys. Rev. A, 2009, 79(2): 025602.

[20] ZHOU L, ZHANG W, LING H Y, et al. Properties of a coupled two-species atom heteronuclear-molecule condensate[J]. Phys. Rev. A, 2007, 75: 043603.

[21] YUROVSKY V A, BEN-REUVEN A, JULIENNE P S, et al. Atom loss and the formation of a molecular Bose-Einstein condensate by Feshbach resonance[J]. Phys. Rev. A, 2000, 62: 043605.

[22] DUINE R A, STOOF H T C. Many-Body Aspects of coherent atom-molecule oscillations[J]. Phys. Rev. Lett., 2003, 91: 150405.

[23] MACKIE M, HÄRKÖNEN K, COLLIN A, et al. Improved efficiency of stimulated Raman adiabatic passage in photoassociation of a Bose-Einstein condensate[J]. Phys. Rev. A, 2004, 70: 013614.

[24] LIU J, LIU B, FU L B. Many-body effects on nonadiabatic Feshbach conversion in bosonic systems[J]. Phys. Rev. A, 2008, 78(1): 013618.

[25] LIU J, FU L B, LIU B, et al. Role of particle interactions in the Feshbach conversion of fermionic atoms to bosonic molecules[J]. New J. Phys., 2008, 10(12): 123018.

[26] PARKINS A, WALLS D F. The physics of trapped dilute-gas Bose-Einstein condensates[J]. Physics Reports, 1998, 303(1): 1-80.

[27] INOUYE S, ANDREWS M, STENGER J, et al. Observation of Feshbach resonance in a Bose-Einstein condensate[J]. Nature, 1998, 392: 151.

[28] DONLEY E A, CLAUSSEN N R, CORNISH S L, et al. Dynamics of collapsing and exploding Bose-Einstein condensates[J]. Nature, 2001, 412: 295-299.

[29] EISENBERG E, HELD K, ALTSHULER B L. Dephasing times in closed quantum dots[J]. Phys. Rev. Lett., 2002, 88: 136801.

[30] KOKKELMANS S J J M F, HOLLAND M J. Ramsey fringes in a Bose-Einstein condensate between atoms and molecules[J]. Phys. Rev. Lett., 2002, 89: 180401.

[31] MACKIE M, SUOMINEN K A, JAVANAINEN J. Mean-field theory of Feshbach-resonant interactions in ^{85}Rb condensates[J]. Phys. Rev. Lett., 2002, 89: 180403.

[32] DOU F Q, LI S C, CAO H, et al. Creating pentamer molecules by generalized stimulated Raman adiabatic passage[J]. Phys. Rev. A, 2012, 85: 023629.

[33] DOU F Q, FU L B, LIU J. Formation of N-body polymer molecules through generalized stimulated Raman adiabatic passage[J]. Phys. Rev. A, 2013, 87(4): 043631.

[34] HUDSON J J, SAUER B E, TARBUTT M R, et al. Measurement of the electron electric dipole moment using YbF molecules[J]. Phys. Rev. Lett., 2002, 89: 023003.

[35] FLAMBAUM V V, KOZLOV M G. Enhanced sensitivity to the time variation of the fine-structure constant and $m_\mathrm{p}/m_\mathrm{e}$ in diatomic molecules[J]. Phys. Rev. Lett., 2007, 99: 150801.

[36] MICHELI A, BRENNEN G, ZOLLER P. A toolbox for lattice-spin models with polar molecules[J]. Nature Phys., 2006, 2(5): 341-347.

[37] RABL P, DEMILLE D, DOYLE J M, et al. Hybrid quantum processors: Molecular ensembles as quantum memory for solid state circuits[J]. Phys. Rev. Lett., 2006, 97: 033003.

[38] CHIN C, KRAEMER T, MARK M, et al. Observation of Feshbach-like resonances in collisions between ultracold molecules[J]. Phys. Rev. Lett., 2005, 94: 123201.

[39] TSCHERBUL T V, KREMS R V. Controlling electronic spin relaxation of cold molecules with electric fields[J]. Phys. Rev. Lett., 2006, 97: 083201.

[40] MACKIE M. Feshbach-stimulated photoproduction of a stable molecular Bose-Einstein condensate[J]. Phys. Rev. A, 2002, 66: 043613.

[41] ALZETTA G, GOZZINI A, MOI L, et al. An experimental method for the observation of r.f. transitions and laser beat resonances in oriented Na vapour[J]. Nuovo Cimento B, 1976, 36: 5-20.

[42] LING H Y, PU H, SEAMAN B. Creating a stable molecular condensate using a generalized Raman adiabatic passage scheme[J]. Phys. Rev. Lett., 2004, 93: 250403.

[43] MANAI I, HORCHANI R, LIGNIER H, et al. Rovibrational cooling of molecules by optical pumping[J]. Phys. Rev. Lett., 2012, 109: 183001.

[44] STUHL B K, HUMMON M T, YEO M, et al. Evaporative cooling of the dipolar hydroxyl radical[J]. Nature, 2012, 492(7429): 396-400.

[45] JONES K M, TIESINGA E, LETT P D, et al. Ultracold photoassociation spectroscopy:

Long-range molecules and atomic scattering[J]. Rev. Mod. Phys., 2006, 78: 483-535.

[46] GIORGINI S, PITAEVSKII L P, STRINGARI S. Theory of ultracold atomic Fermi gases[J]. Rev. Mod. Phys., 2008, 80: 1215-1274.

[47] DE MELO C A S. When fermions become bosons: Pairing in ultracold gases[J]. Physics Today, 2008, 61: 45.

[48] FRIEDRICH B, DOYLE J M. Why are cold molecules so hot?[J]. ChemPhysChem, 2009, 10(4): 604-623.

[49] SHUMAN E, BARRY J, DEMILLE D. Laser cooling of a diatomic molecule[J]. Nature, 2010, 467(7317): 820-823.

[50] KÜPPER J, FILSINGER F, MEIJER G. Manipulating the motion of large neutral molecules[J]. Faraday Discuss., 2009, 142: 155-173.

[51] PATTERSON D, TSIKATA E, DOYLE J M. Cooling and collisions of large gas phase molecules[J]. Phys. Chem. Chem. Phys., 2010, 12(33): 9736-9741.

[52] SABBAH H, BIENNIER L, SIMS I R, et al. Understanding reactivity at very low temperatures: The reactions of oxygen atoms with alkenes[J]. Science, 2007, 317(5834): 102-105.

[53] HORNBERGER K, UTTENTHALER S, BREZGER B, et al. Collisional decoherence observed in matter wave interferometry[J]. Phys. Rev. Lett., 2003, 90: 160401.

[54] BARTELS R A, WEINACHT T C, WAGNER N, et al. Phase modulation of ultrashort light pulses using molecular rotational wave packets[J]. Phys. Rev. Lett., 2001, 88: 013903.

[55] ZEPPENFELD M, ENGLERT B G E, GLÖCKNER R, et al. Sisyphus cooling of electrically trapped polyatomic molecules[J]. Nature, 2012, 491(7425): 570-573.

[56] BARRY J F, DEMILLE D. Low-temperature physics: A chilling effect for molecules[J]. Nature, 2012, 491: 539-540.

[57] CAMPBELL W C, TSCHERBUL T V, LU H I, et al. Mechanism of collisional spin relaxation in $^3\Sigma$ molecules[J]. Phys. Rev. Lett., 2009, 102: 013003.

[58] HUDSON E R, TICKNOR C, SAWYER B C, et al. Production of cold formaldehyde molecules for study and control of chemical reaction dynamics with hydroxyl radicals[J]. Phys. Rev. A, 2006, 73: 063404.

[59] HUMMON M T, TSCHERBUL T V, KLOS J, et al. Cold N+NH collisions in a magnetic trap[J]. Phys. Rev. Lett., 2011, 106: 053201.

[60] WALLIS A O G, HUTSON J M. Production of ultracold NH molecules by sympathetic cooling with Mg[J]. Phys. Rev. Lett., 2009, 103: 183201.

[61] TSCHERBUL T V, YU H G, DALGARNO A. Sympathetic cooling of polyatomic molecules with S-state atoms in a magnetic trap[J]. Phys. Rev. Lett., 2011, 106: 073201.

[62] WEINSTEINAND J D, DECARVALHO R, GUILLET T, et al. Magnetic trapping of calcium monohydride molecules at millikelvin temperatures[J]. Nature, 1998, 395(6698): 148-150.

[63] FULTON R, BISHOP A I, BARKER P. Optical Stark decelerator for molecules[J]. Phys. Rev. Lett., 2004, 93(24): 243004.

[64] JING H, CHENG J, MEYSTRE P. Coherent atom-trimer conversion in a repulsive Bose-Einstein condensate[J]. Phys. Rev. Lett., 2007, 99: 133002.

[65] JING H, CHENG J, MEYSTRE P. Coherent generation of triatomic molecules from ultracold atoms[J]. Phys. Rev. A, 2008, 77: 043614.

[66] EFIMOV V. Energy levels arising from resonant two-body forces in a three-body system[J]. Phys. Lett. B, 1970, 33(8): 563-564.

[67] KRAEMER T, MARK M, WALDBURGER P, et al. Evidence for Efimov quantum states in an ultracold gas of caesium atoms[J]. Nature, 2006, 440(7082): 315-318.

[68] BRAATEN E, HAMMER H W. Efimov physics in cold atoms[J]. Ann. Phys., 2007, 322(1): 120-163.

[69] KNOOP S, FERLAINO F, MARK M, et al. Observation of an Efimov-like trimer resonance in ultracold atom–dimer scattering[J]. Nature Phys., 2009, 5(3): 227-230.

[70] MACHTEY O, SHOTAN Z, GROSS N, et al. Association of Efimov trimers from a three-atom continuum[J]. Phys. Rev. Lett., 2012, 108: 210406.

[71] STOLL M, KÖHLER T. Production of three-body Efimov molecules in an optical lattice[J]. Phys. Rev. A, 2005, 72: 022714.

[72] HAMMER H W, PLATTER L. Universal properties of the four-body system with large scattering length[J]. Eur. Phys. J. A-Hadrons and Nuclei, 2007, 32(1): 113-120.

[73] VON STECHER J, D'INCAO J P, GREENE C H. Signatures of universal four-body phenomena and their relation to the Efimov effect[J]. Nature Phys., 2009, 5(6): 417-421.

[74] POLLACK S E, DRIES D, HULET R G. Universality in three-and four-body bound states of ultracold atoms[J]. Science, 2009, 326(5960): 1683-1685.

[75] SCHMIDT R, MOROZ S. Renormalization-group study of the four-body problem[J]. Phys. Rev. A, 2010, 81: 052709.

[76] YAMASHITA M T, FEDOROV D V, JENSEN A S. Universality of Brunnian (N-body Borromean) four- and five-body systems[J]. Phys. Rev. A, 2010, 81: 063607.

[77] HADIZADEH M R, YAMASHITA M T, TOMIO L, et al. Scaling properties of universal tetramers[J]. Phys. Rev. Lett., 2011, 107: 135304.

[78] FERLAINO F, KNOOP S, BERNINGER M, et al. Evidence for universal four-body states tied to an Efimov trimer[J]. Phys. Rev. Lett., 2009, 102: 140401.

[79] THØGERSEN M, FEDOROV D V, JENSEN A S. N-body Efimov states of trapped bosons[J]. Europhys. Lett., 2008, 83(3): 30012.

[80] VON STECHER J. Five- and six-body resonances tied to an Efimov trimer[J]. Phys. Rev. Lett., 2011, 107: 200402.

[81] VON STECHER J. Weakly bound cluster states of Efimov character[J]. J. Phys. B: Atomic, Molecular and Optical Physics, 2010, 43(10): 101002.

[82] HANNA G J, BLUME D. Energetics and structural properties of three-dimensional bosonic clusters near threshold[J]. Phys. Rev. A, 2006, 74: 063604.

[83] JING H, JIANG Y. Coherent atom-tetramer conversion: Bright-state versus dark-state

schemes[J]. Phys. Rev. A, 2008, 77: 065601.

[84] LI G Q, PENG P. Formation of a heteronuclear tetramer A_3B via Efimov-resonance-assisted stimulated Raman adiabatic passage[J]. Phys. Rev. A, 2011, 83: 043605.

[85] PARKINS A, WALLS D. The physics of trapped dilute-gas Bose-Einstein condensates[J]. Phys. Rep., 1998, 303(1): 1-80.

[86] LING H Y, MAENNER P, ZHANG W, et al. Adiabatic theorem for a condensate system in an atom-molecule dark state[J]. Phys. Rev. A, 2007, 75: 033615.

[87] WIDERA A, MANDEL O, GREINER M, et al. Entanglement interferometry for precision measurement of atomic scattering properties[J]. Phys. Rev. Lett., 2004, 92: 160406.

[88] MENG S Y, FU L B, LIU J. Adiabatic fidelity for atom-molecule conversion in a nonlinear three-level Λ system[J]. Phys. Rev. A, 2008, 78(5): 053410.

[89] LU L H, LI Y Q. Atom-to-molecule conversion efficiency and adiabatic fidelity[J]. Phys. Rev. A, 2008, 77: 053611.

[90] DULIEU O, GABBANINI C. The formation and interactions of cold and ultracold molecules: new challenges for interdisciplinary physics[J]. Rep. Prog. Phys., 2009, 72(8): 086401.

[91] KUZNETSOVA E, PELLEGRINI P, CÔTÉ R, et al. Formation of deeply bound molecules via chainwise adiabatic passage[J]. Phys. Rev. A, 2008, 78: 021402.

[92] THALHAMMER G, WINKLER K, LANG F, et al. Long-lived Feshbach molecules in a three-dimensional optical lattice[J]. Phys. Rev. Lett., 2006, 96: 050402.

[93] VOLZ T, SYASSEN N, BAUER D M, et al. Preparation of a quantum state with one molecule at each site of an optical lattice[J]. Nature Phys., 2006, 2(10): 692-695.

[94] WUNSCH B, ZINNER N T, MEKHOV I B, et al. Few-body bound states in dipolar gases and their detection[J]. Phys. Rev. Lett., 2011, 107: 073201.

[95] ZINNER N T, WUNSCH B, MEKHOV I B, et al. Few-body bound complexes in one-dimensional dipolar gases and nondestructive optical detection[J]. Phys. Rev. A, 2011, 84: 063606.

[96] 赵岫鸟, 孙建安, 豆福全. 外场形式对超冷原子-多聚物分子转化效率的影响 [J]. 物理学报, 2014, 63(22): 220302.

[97] BARTENSTEIN M, ALTMEYER A, RIEDL S, et al. Precise determination of ^{6}Li cold collision parameters by radio-frequency spectroscopy on weakly bound molecules[J]. Phys. Rev. Lett., 2005, 94: 103201.

[98] PU H, MAENNER P, ZHANG W, et al. Adiabatic condition for nonlinear systems[J]. Phys. Rev. Lett., 2007, 98: 050406.

[99] MENG S Y, CHEN X H, WU W, et al. Instability, adiabaticity, and controlling effects of external fields for the dark state in a homonuclear atom-tetramer conversion system[J]. Chin. Phys. B, 2014, 23(4): 040306.

[100] WINKLER K, THALHAMMER G, THEIS M, et al. Atom-molecule dark states in a Bose-Einstein condensate[J]. Phys. Rev. Lett., 2005, 95: 063202.

[101] Drummond P D, Kheruntsyan K V, Heinzen D J, et al. Stimulated Raman adiabatic passage

from an atomic to a molecular Bose-Einstein condensate[J]. Phys. Rev. A, 2002, 65: 063619.

[102] IVANOV P, VITANOV N, BERGMANN K. Effect of dephasing on stimulated Raman adiabatic passage[J]. Phys. Rev. A, 2004, 70(6): 063409.

[103] 孟少英, 刘杰. 超冷原子–分子转化动力学: 受激拉曼绝热过程 [J]. 物理学进展, 2010, 30(3): 280-295.

[104] MENG S Y, FU L B, CHEN J, et al. Linear instability and adiabatic fidelity for the dark state in a nonlinear atom-trimer conversion system[J]. Phys. Rev. A, 2009, 79(6): 063415.

[105] NAIDON P, ENDO S. Efimov physics: A review[J]. Rep. Prog. Phys., 2017, 80(5): 056001.

编 后 记

  《博士后文库》(以下简称《文库》)是汇集自然科学领域博士后研究人员优秀学术成果的系列丛书.《文库》致力于打造专属于博士后学术创新的旗舰品牌,营造博士后百花齐放的学术氛围,提升博士后优秀成果的学术和社会影响力.

  《文库》出版资助工作开展以来,得到了全国博士后管委会办公室、中国博士后科学基金会、中国科学院、科学出版社等有关单位领导的大力支持,众多热心博士后事业的专家学者给予积极的建议,工作人员做了大量艰苦细致的工作.在此,我们一并表示感谢!

《博士后文库》编委会